T0135767

Martin Guski

Influences of external error sources
on measurements of room acoustic parameters

Logos Verlag Berlin GmbH

Aachener Beiträge zur Technischen Akustik

Editor:
Prof. Dr. rer. nat. Michael Vorländer
Institute of Technical Acoustics
RWTH Aachen University
52056 Aachen
www.akustik.rwth-aachen.de

Bibliographic information published by the Deutsche Nationalbibliothek

The Deutsche Nationalbibliothek lists this publication in the Deutsche Nationalbibliografie; detailed bibliographic data are available in the Internet at http://dnb.d-nb.de .

D 82 (Diss. RWTH Aachen University, 2015)

ISBN 978-3-8325-4146-0
ISSN 1866-3052
Vol. 24

Logos Verlag Berlin GmbH
Comeniushof, Gubener Str. 47,
D-10243 Berlin
Tel.:   +49 (0)30 / 42 85 10 90
Fax: +49 (0)30 / 42 85 10 92
http://www.logos-verlag.de

INFLUENCES OF EXTERNAL ERROR SOURCES ON MEASUREMENTS OF ROOM ACOUSTIC PARAMETERS

Von der Fakultät für Elektrotechnik und Informationstechnik der
Rheinisch-Westfälischen Technischen Hochschule Aachen
zur Erlangung des akademischen Grades eines

DOKTORS DER INGENIEURWISSENSCHAFTEN

genehmigte Dissertation

vorgelegt von

Dipl.-Ing.

Martin Guski

aus Tönisvorst

Berichter:

Universitätsprofessor Dr. rer. nat. Michael Vorländer
Universitätsprofessor Dr.-Ing. Dirk Heberling

Tag der mündlichen Prüfung: 26. Oktober 2015

Diese Dissertation ist auf den Internetseiten der Hochschulbibliothek online
verfügbar.

meinen Eltern gewidmet

Abstract

Correct and precise measurements of room acoustic parameters are of fundamental importance for subjective room impression characterization and for the physical description of the sound field. This thesis investigates external influences on acoustic measurements and errors of the resulting room acoustic parameters. Theoretical models have been developed to predict these errors and indicate the tolerable limits of the described influences. To validate these models, specially designed room acoustic measurements that separate the individual influence factors have been conducted.

Noise is one of the main influence factors and occurs during every measurement. In this thesis the performance of five commonly used stationary noise compensation methods are systematically analyzed depending on the peak signal-to-noise ratio. A theoretical model has been developed that demonstrates that these five techniques show clear differences regarding the size of the error. Long-term measurements that clearly validate the model have been conducted. Impulsive noise that could also occur during the measurement is investigated separately, as the previously introduced compensation techniques are unsuited to handle the influence. It is shown that even for impulsive noise events that are below the level of the excitation signal, the error in the evaluated room acoustic parameter is large. To avoid this additional error source, this thesis describes an automatic algorithm to detect the occurrence of impulsive noise in a measured impulse response.

The second part of this thesis analyzes the influence of time variances during measurements: temperature changes, air movement, and scattering objects. Investigations of temperature changes during one measurement show a significant influence on the parameters. The sensitivity of the evaluated reverberation time is clearly higher than the theoretical prediction. The tolerable differences in temperature for different measurements have also been investigated in further measurements. The results demonstrate that the theoretical assumptions underestimate the error and that theory can therefore not be used to predict or compensate this effect. The determined deviations for inter- and intra-measurement changes are documented to allow error estimation. The studies of human-sized scattering objects show only an influence on the clarity index nearby the object. Air movements caused by an activated ventilation system have been found to increase the random error of room acoustic parameters significantly.

However, the magnitude of the error is low and can be neglected for conventional measurements.

This thesis shows that especially noise and temperature changes during the measurements can introduce significant errors in the room acoustic parameters and therefore have to be given particular consideration.

Contents

1

Introduction

The requirements for the acoustic conditions of a room are strongly dependent on the intended type of usage. Concert halls or opera houses, for example, are supposed to have long reverberation to support the music performed. For conference halls or rooms for spoken performances, on the other hand, a long reverberation cause smearing of subsequent syllables and is therefore undesirable. Besides the reverberation, there are further details of the sound field (for example early reflections or angle of incidence) that have an influence on the human perception.

Several room acoustic parameters have been developed to quantitatively determine several perceptual aspects. Therefore, room acoustic measurements are conducted to obtain the room impulse response (RIR) and the parameters can be derived therefrom.

All acoustics properties of a room for a given source and receiver position are characterized by the RIR, which represents the answer of the room to a very short and loud impulse, such as a handclap, or a pistol shot. The impulse propagates in numerous ways throughout the room and, depending on the length of the sound path and the reflections on the walls, a characteristic RIR is formed. Generally, the density of the reflections increases for the later parts of the impulse response and the amplitudes decrease.

Figure 1.1 shows an example of a room impulse response. At the beginning, the direct sound and early reflections can be distinguished. Because of the large time difference between the early reflections, they can be seen as discrete single events. In the later part of the RIR, the density of the reflections increases, and

1

they can only be described as statistical reverberance.

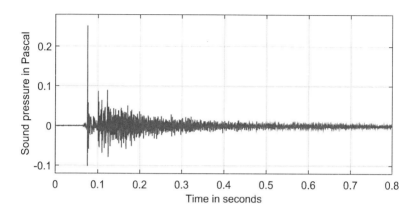

Figure 1.1.: The measured room impulse response describes the acoustic conditions completely. Several room acoustic parameters can be derived directly from the room impulse response.

The reverberation time is the oldest and one of the most important room acoustic parameters. It is determined from the later part of the impulse response and describes the average decay of the sound power. The early reflections cannot be perceived individually by humans and contribute in this way to the sensed loudness of the direct sound [1, 2]. Reflections that arrive later can be perceived as echoes and can therefore interfere with the direct sound. Room acoustic parameters, such as the clarity index, model these perceptual effects by determining the energetic ratio of early and late reflections. The precision of these measured parameters is important to compare different auditoria, verify reconstruction measures, or contractually defined requirements.

The measured reverberation time is also of fundamental importance in describing the physical sound field. It is further essential for numerous acoustic measurement procedures, such as for example, sound insulation, sound power, absorption, and the scattering coefficient. The precision of these derived parameters directly depends on the accuracy of the reverberation time.

The measurements of room acoustic parameters have been standardized in ISO

3382-1 [3] "Acoustics - Measurement of room acoustic parameters" to improve the quality and comparability of these parameters. The largest error sources are avoided by defining calculation algorithms, measurement equipment, and conditions. Nevertheless, there are undefined details in the complex measurement process, and some rules are rather lenient to ensure a backwards compatibility of old measurement equipment.

ISO 3382 also summarizes the current state of research concerning the just noticeable differences (JND) for the defined parameters. The JNDs (also known as difference limens) have been determined in elaborate listening tests and define the change in a room acoustic parameter that can be perceived by humans. The difference limens are important in the discussion of error sizes, since they define the maximum tolerable deviation for measurements.

Due to the importance of this parameter, several measurements were performed and discussed in literature. In 1992, Pelorson, Vian, and Polack [4] tested the influence of different elements of the measuring system and found large impacts on room acoustic parameters. Lundeby et al. [5] conducted a measurement round robin and compared the variances of the results. Seven different teams, each equipped with own measurement hardware and evaluation software, measured the same room according to ISO 3382. The standard deviations of several investigated parameters were in the range or a little higher than the JNDs.

Bradley [6] organized another round robin where a digital reverberator device was shipped to different measurement groups in the world. Due to the use of the digital device, a number of parts of the measurement chain were not contained in the analysis. Errors that usually occur due to loudspeaker, microphone, and the real room are excluded. This concerns, for example, directivity, frequency responses, and nonlinearities of loudspeakers and microphones. External influences such as time-variances that occur in the room and the major noise source (ambient noise) were also excluded in this investigation. A number of 23 different room acoustic measurement systems were compared. Since the correct room acoustic parameters of the digital device were unknown, the mean value is considered as best estimate and the standard deviation of all results have been interpreted. The observed deviations were small compared to the human difference limens, implying robustness of the investigated parts of the measurement process.

Katz [7] excluded the complete measurement hardware from his round robin and provided a digital impulse response to 37 different organizations. This way, the study was independent of the unequal hardware of the participants and focused on the signal-processing algorithms. In contrast to Bradley et al., the impulse response used was a real measurement and therefore contained noise. Because the measurement has been conducted using a bursting balloon, the noise level was relatively high, leading to clear differences in the results. Katz found variations larger than the JND for the lower frequencies that could be lead back to the stationary noise floor detection and noise compensation techniques.

1.1. Objective

The objective of this thesis is to study single external effects that influence acoustic measurements and investigate their impact on room acoustic parameters in detail. Therefore, the underlying physical effects are used to develop models that are able to predict the error impact on the parameters. These models can be used to clearly identify the existing relationships and allow a prediction of the expected error, based on the circumstances of a measurement. It is also possible to determine the tolerable magnitude of the disturbing influence so as to stay below a defined error of the parameter.

It is of fundamental importance to validate the developed models, because the theoretical assumptions are often strong simplifications of reality. It has to be ensured that the neglected details have an insignificant impact on the parameters and no ignored effects are present. Therefore, special measurement setups are designed that modify only the investigated influence and keep the remaining conditions as constant as possible. For influences that cannot be modeled, the measurements provide empirical data of the error, which can be used to assess the size of the error.

1.2. Outline

Theoretical fundamentals such as measurement theory, room acoustic theory, and parameters are summarized in Chapter 2. The effects of stationary noise and the performance of the compensation techniques are investigated in Chapter 3. Impulsive noise is an exception of the stationary assumption. The effects are discussed in detail in Chapter 4, and an automatic noise compensation algorithm is presented to overcome this problem.

Chapter 5 summarizes the analysis of time variances that occur during measurements. The influences of air movement, temperature changes, and human scattering objects are investigated separately. The determination of the impact of temperature changes is distinguished between changes that occur between two single measurements and within one measurement.

2

Fundamentals

This chapter gives a brief introduction into the fundamentals that are used in this thesis. Basic concepts such as measurement theory, room acoustic theory, and room acoustic parameter analysis are covered.

2.1. Measurement Theory

The commonly used system theory approach assumes a linear and time-invariant (LTI) system. This allows describing every system by its complex valued and frequency dependent transfer function $H(f)$.

In this context, linearity means that the output of the system, which is here determined as $y(t)$, only contains the same frequencies that are also present in the input signal $x(t)$. The amplitude of every single frequency f is scaled by the factor $|H(f)|$ and the frequency is shifted by the phase of the transfer function $\arg(H(f))$. Time-invariance means that the behavior is constant over time. If the system response to the input signal $x(t)$ is $y(t)$, an arbitrary time shift of the input signal $x(t-T)$ will result in a time shift of the output signal $y(t-T)$. Further information concerning LTI systems can be found in Oppenheim, Schafer, and Buck [8] and Ohm and Lüke [9].

The transfer function of an LTI system can be calculated by dividing the output by the input for every single frequency:[1]

$$H(f) = \frac{Y(f)}{X(f)}.$$ (2.1)

Equivalently to this frequency domain representation, the system can be described in the time domain by the impulse response (IR) $h(t)$. The impulse response is the answer at the output of the system to an infinitely high and infinitely short Dirac impulse $\delta(t)$ at the input. Both representations of the system (impulse response and transfer function) can be converted into each other using the Fourier transform:

$$H(f) = \int_{-\infty}^{\infty} h(t) \cdot \exp\left(-j2\pi ft\right) \, \mathrm{d}t$$ (2.2)

$$h(t) = \int_{-\infty}^{\infty} H(f) \cdot \exp\left(j2\pi ft\right) \, \mathrm{d}f \, .$$ (2.3)

Because of this mathematical relation, the impulse response and the transfer function are of equal importance and the terms are often used conterminously in this thesis.

Nowadays, the complete signal processing is digital and therefore the signals are presented in a sampled and quantized form. This introduces a few minor changes in the presented theory. These changes are not discussed into further detail at this point, but are take up if relevant.

2.1.1. Room Acoustic Measurements

The first room acoustic measurements were performed by W.C. Sabine in 1922. He used organ pipes to excite a stationary sound field in a room and measured the time needed for the sound field to decay. In this first approach he used his

---

[1]Initially it is assumed that $X(f) \neq 0$

ears to detect the level decay and defined the reverberation time. Over time, technical progress, with availability of electro-acoustic transducers, electronic recording devices, and the personal computer changed the measurement procedure and increased the quality of the measurements. Nevertheless, the basic principle of Sabine is still a valid measurement method according to the room acoustic measurement standard ISO 3382-1 [3], if a loudspeaker for excitation, a microphone, and a level recorder are used.

However, the commonly used method to carry out room acoustic measurements is the impulse response method. The impulse response completely describes the acoustic properties of the room from the point of the sound source to the receiver position. Based on this room impulse response (RIR) a variety of room acoustic parameters (including the aforementioned reverberation time, see Chapter 2.3.1) can be calculated. Furthermore, the RIR can be used to simulate the received signal for any given input signal played back at the source position (also known as auralization [10]).

The simplest way to determine the RIR is to excite the room with a pistol, bursting balloon, or hand clapper. The impulse response can directly be recorded with a microphone at the receiver position. The implementation is very easy, the procedure is fast and the needed equipment is minimal. However, the disadvantages of this method are poor reproducibility, a deviation of the often-required omnidirectional directivity pattern, and the low ratio between signal and ambient noise.

Because of these disadvantages, typically an electro-acoustic excitation is preferred. In combination with a personal computer, different excitation signals can be used that significantly increase the measurement in many aspects. Therefore, in this thesis only the impulse response method with loudspeakers is considered. In Subsection 2.1.2, the measurement chain is described and in Subsection 2.1.3, different excitation signals are discussed.

2.1.2. Measurement Chain

This chapter gives a brief overview of all devices involved in a typical room acoustic measurement. The central device is a personal computer (PC) with adequate software to execute the measurement and perform the necessary post-processing. In this thesis a normal desktop PC with Matlab and the ITA-Toolbox [11] is used.

A schematic overview of the measurement chain can be seen in Figure 2.1. The PC generates (or stores) the digital excitation signal. The digital-to-analog (DA) converter transforms the digitally sampled and quantized signal into an analog continuous electrical signal. As indicated by the name, the amplifier amplifies the low power DA output signal to drive a transducer. The loudspeaker transforms the electrical signal into an acoustic signal. According to the commonly used standard for room acoustic measurements the "sound source shall be as close to omnidirectional as possible" [3]. However, ISO 3382 is not very strict, since the limit values have to be measured only at one elevation angle and is averaged over 30 degrees in azimuth. The standard also requires a sufficiently high sound pressure level to obtain the required decay range (see Section 2.3). The measured device under test is the room itself and the loudspeaker is usually placed on a position where the sound source (orchestra, musician or speaker) is typically located.

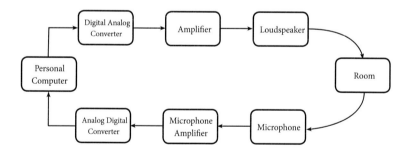

Figure 2.1.: Block diagram of the elements typically used in measurement chain for room acoustic measurements.

Consequently, the microphones are located at typical audience positions. Microphones transform the acoustic signal back into an electrical signal. The directivity depends on the purpose of the measurement or on the parameter to be evaluated. Most parameters are based on measurements with omnidirectional receivers. Some parameters that describe the spatial impression need special microphones with figure-of-eight characteristic to separate the lateral components of the received signal. Other spatial parameters need a detailed approximation of the directivity of a human. Therefore, models of human heads (often including torso) that are equipped with two microphones in their ears are used. These binaural measurements can be used for auralization to enable a realistic spatial hearing for the listener. The electric output signal of the microphone is amplified by the microphone amplifier and the analog digital (AD) converter samples and quantizes the signal and provides the digital recorded signal. AD and DA converters usually belong to the same sound card to ensure a synchronized sampling.

The described measurement chain can be implemented with multiple channels for input and output. Several input channels that enable measuring more than one microphone position simultaneously are often used to reduce the total time of the measurement session. For the output chain often two or three channels are used to drive multi-way speakers. Further requirements for every part of the measurement chain are linearity, time-invariance, spectral flatness, and low noise level.

As mentioned before, the used measurement theory is based on LTI systems. While high-quality amplifiers and sound cards can be described in a good approximation by LTI systems, the loudspeaker can show nonlinear components. Since the nonlinear components increase with higher sound pressure levels, it is a major advantage to use excitation signals that indicate the magnitude of these distortions (i.e. exponential sweeps, see Section 2.1.3).

Every part of the measurement chain introduces noise into the measurement. However, the magnitudes of the noise contributions differ strongly. Typically, the greatest contribution is the inevitable ambient noise in the room that is recorded by the microphone. It is caused by various sources inside and outside the room and has a typical frequency dependent power density spectrum that decreases with frequency by a factor of $1/f$ (also known as pink spectrum). Since it is also recorded by the microphone, the ambient noise is also reinforced by the microphone amplifier. One smaller contribution is the electronic noise that is introduced by every electric device and has various causing effects, such as semiconductor noise or thermal noise.

Utilized Measurement Equipment

This subsection describes the equipment used for the measurements in this thesis. The sound card is a *Digiface* by *RME* in combination with three *RME Octamic II* front ends. The *Octamic II* contains eight input channels, each equipped with a 24 bit AD converter and a high-class microphone pre-amplifier. The loudspeaker amplifier is an *HK Audio VC 2400*, a two-channel amplifier with a total power of 2400 W. Two types of microphones are used for the measurements: A *Brüel & Kjær* $\frac{1}{2}$-inch free field microphone model *Type 4190*. In addition, a *Nexus* Charge Amplifier of *Type 2690* is used to provide the supply voltage for the B&K microphone. Up to 24 *Sennheiser KE-4* microphones with custom-made impedance transformers are utilized to allow simultaneous measurements of various positions.

The dodecahedron loudspeaker used for the measurements was designed and constructed by the Institute of Technical Acoustics in Aachen, Germany. It was developed to obtain an omnidirectional directivity and a flat frequency response over a wide frequency range to measure impulse responses suitable for auralization purposes. Since it is not possible to achieve an omnidirectional radiation over the full audible frequency range with one single cabinet, the loudspeaker is realized using a three band design (see Figure 2.2).

The core of the subwoofer is a 12-inch driver. The cabinet has a radiation opening of only 10 cm to ensure an omnidirectional radiation up to 300 Hz. The tuning of the housing results in a flat frequency response from 40 Hz to 300 Hz (-6 dB). The medium frequency range is covered by a spherical cabinet with a diameter of 30 cm. It consists of 12 loudspeakers, each 12 cm in diameter, and is placed 25 cm above the opening of the woofer. The usable frequency range is up to 6.3 kHz and the first deviation of directivity above 3 dB occurs at 2000 Hz. However, the deviation limits according to ISO 3382 are defined on a smoothed version of the directivity. The maximum deviation of the dodecahedron after averaging is 1.3 dB in the 4 kHz octave band and 3.5 dB for 8 kHz. ISO 3382 specifies the

Figure 2.2.: Three way omnidirectional dodecahedron loudspeaker used for the measurements.

maximum acceptable deviation in the 4 kHz octave band as ±6 dB and gives no limits for 8 kHz. Figure 2.3 shows the averaged directivity of the mid-range speaker. It can be seen that the speaker is always far below the limits, even for high frequencies.

The high frequency unit is sphere made of aluminum with a diameter of 9.5 cm. It is equipped with twelve tweeters of 0.8 inch diameter. The directivity can be considered as omnidirectional for a frequency up to 5 kHz, considering a ± 3 dB limit of the directivity without smoothing.

2.1.3. Excitation Signal

The major benefit of using loudspeakers is the flexibility concerning the excitation signal. The properties of the signal have crucial influence on the quality of the result. In the following section, a selection of the most important and commonly used signals is described.

13

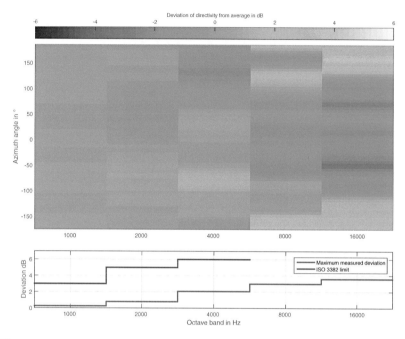

Figure 2.3.: Directivity of the mid-range speaker averaged according to ISO 3382 (top). The maximum deviation of the directivity from the average is compared with the limits defined in ISO 3382 in the lower part of the figure. The mid-range speaker remain below the limits even for frequencies far above its frequency range.

Impulse

The most obvious excitation signal is a Dirac impulse, since this corresponds to the definition of an impulse response. The advantage is that no post-processing is necessary, since the recorded signal is already the wanted impulse response. A Dirac impulse is defined as an impulse with infinitely high amplitude and a duration approaching zero:

$$\delta(t) = \begin{cases} \infty & \text{for } t = 0 \\ 0 & \text{else} \end{cases} . \tag{2.4}$$

The equivalent time-discrete signal, the Kronecker delta, already shows the physical limitations with duration of one sample (after DA conversion) and amplitude of one (maximum in digital domain):

$$\delta_k(n) = \begin{cases} 1 & \text{for } n = 0 \\ 0 & \text{else} \end{cases} . \tag{2.5}$$

Despite the amplification in the output measurement chain, for normal room acoustic measurements the emitted energy of the signal will be not much higher than the noise energy. Additionally to the resulting low signal-to-noise ratio (SNR), a further disadvantage is the high excitation on nonlinearities. Due to the short high amplitude, the loudspeaker is in a critical operating point, typically causing nonlinear behavior.

Maximum Length Sequences

Maximum length sequences (MLS) are pseudorandom binary sequences that can be utilized as excitation signals. The main advantage of using a longer signal is that more signal energy is available, resulting in a significantly better SNR. A feedback shift register of length m can be used to generate MLS of length $l = 2^m - 1$. The spectrum of the MLS signal is white and the autocorrelation function is approximately a Dirac impulse. This fact allows calculating the impulse response as the cross-correlation of the recorded maximum length sequence and the excitation signal.

An advantage of the MLS technique and the main reason for its popularity is the possibility to use the fast Hadamard transform to calculate the correlation function [12, 13]. Compared to other techniques, the computational requirements are significantly smaller, which was an important factor in the 80s and 90s. Since nowadays the computational power has increased drastically, the former advantage has become insignificant.

A disadvantage of the maximum length sequences is the sensitivity to nonlinear distortions. For MLS measurements the distortions cause spikes spread over the entire impulse response. It is not possible to remove the nonlinear components. The spikes first appear in the noise part of the impulse response, because of the lower amplitudes, but with increasing nonlinearities also the signal part is affected, making the complete measurement useless [14].

Another advantage is the robustness of MLS against impulsive noise during the measurement. The impulsive noise is transformed into a stationary noise and thus distributed over the entire impulse response. Due to this time spreading, the noise amplitude is considerably reduced. The random temporal structure and the low amplitudes allow considering impulsive noise as an additional stationary noise term in the impulse response as well. All noise detection and compensation techniques are thus taken into account in a similar way.

Sweeps

A sweep (also known as chirp or swept sine) is a signal with constantly increasing frequency. The analytic formulation of sweeps has the advantage of freely selectable bandwidth and length of the excitation signal. Usually, a so-called stop margin is flowed by the actual signal. In this silent part, no signal is played back while the recording is continued. This ensures that all frequencies have enough time to decay before the recording stops.

Sweeps offer several advantages compared to other excitation signals such as MLS or random noise [15, 16]. Compared to MLS measurements, sweeps are more robust to time variances. Since for sweeps every frequency is excited during a relatively short moment, time variances in the system (for example, air movement in the room or temperature changes of an amplifier, transducer, or the room) result only in small errors. For MLS and noise measurements, all frequencies are excited (pseudo)randomly in the entire measurement time, and the resulting transfer function is an average of these conditions. Therefore, the resulting impulse response might include inaccuracies caused by superposition effects of those variances [17].

Deconvolution Compared to the aforementioned signals, the post-processing differs: The recorded sweep at the microphone position has to be divided by the excitation signal in the frequency domain to obtain the impulse response. Compared to the MLS technique, the computational effort is larger due to multiple Fourier transforms. However, this difference is hardly noticeable with modern computers.

For sweeps that do not cover the full bandwidth, the parts of the spectrum outside this bandwidth contain little energy. Due to the deconvolution, the recorded noise in these frequency parts are amplified very strongly, caused by the division of small values, and this significantly compromises the result. This inversion problem can be solved by using the regularization technique proposed by Kirkeby [18]. A frequency-dependent regularization parameter is adjusted to the relevant frequency range to limit the lowest values of the spectrum before inversion.

One main disadvantage of sweeps is a result of the deconvolution process: the high sensitivity to impulsive noise during the measurement. The detailed problematic and possible solutions are described in Chapter 4.

The discrete Fourier transform (DFT) assumes periodic time signals. This way a so-called cyclic (de-)convolution is used if both spectra are divided. All signal parts that normally would appear at times $t < 0$ appear at the end of the impulse response. This behavior can be avoided by using linear (de-)convolution. However, linear convolution requires a two to three time longer recording time or - for the case of zero padding is used - results in non stationary noise conditions. For that reason the cyclic convolution is preferred and used in this thesis. The length of the sweep is chosen to be at least twice as long as the visible decay of the room to ensure that effects caused by the cyclic convolution do not overlap with the room impulse response.

There is a variety of types of sweeps. The most important ones are discussed below.

Linear Sweep A sweep with constant envelope and a linear increasing frequency is called linear sweep:

$$s_{lin}(t) = \sin\left(2\pi\left(f_1 + \frac{f_2 - f_1}{t_{\mathrm{IR}}}t\right)\right) \ . \tag{2.6}$$

This sweep covers a frequency range from f_1 to f_2 in a total duration of the signal t_{IR}. Since every frequency is excited by the same amount of time, the spectral power density of a linear sweep is constant over frequency (also called white following the spectrum of light).

Exponential Sweep The most common type is the exponential sweep (sometimes also referred to as logarithmic sweep) [14]:

$$s(t) = \sin\left(2\pi f_1\left(\exp\left(\frac{t}{L}\right) - 1\right)\right) \ . \tag{2.7}$$

Here L defines the exponentially increasing frequency as $L = t_{\mathrm{IR}}/\log(\frac{f_2}{f_1})$ with time of signal t_{IR} and bandwidth defined by f_1 and f_2. The exponentially increasing frequency results in a decreasing power density spectrum with 3 dB per octave (called pink spectrum). Since most noise sources in acoustics also show a power spectrum density inversely proportional to frequency, this is an advantage because the signal to noise ratio becomes approximately frequency independent.

Another benefit of exponential sweeps is the behavior of systems with nonlinear components [19]. Due to the exponential dependence of instantaneous frequency on time, every harmonic component appears as a time-shifted version of the fundamental sweep. The time shift increases with the order of the harmonic and depends on the parameters of the sweep. The division of the excitation signal causes all components of a certain harmonic component to be transformed to one point of time in the resulting impulse response. The "harmonic impulse responses" appear before the fundamental impulse response, with the same time shift as harmonic sweeps in the recorded signal. In case of cyclic division, all components of the impulse response appearing at time $t < 0$ are cyclically wrapped to the end of the signal. In Figure 2.4 the nonlinear components can be seen at the end of the RIR.

The clear appearance of nonlinearities allows a fast and objective estimation of the nonlinear components of the system compared to the fundamental components. This allows controlling nonlinear components precisely, for example, by regulating the output amplification. Other excitation signals do not provide this advantage: For linear sweeps the time in the RIR for one harmonic component is dependent on the frequency. This results in a smearing of the harmonic impulse response and it complicates the estimation of the magnitude of the nonlinearities.

Another advantage of the distinct appearance of the nonlinear components for exponential sweeps is the possibility to remove these components in the post-processing by simple truncation of the RIR. However, Ćirić et al. have shown that the nonlinearities also introduce an error to the fundamental part of the impulse response [20].

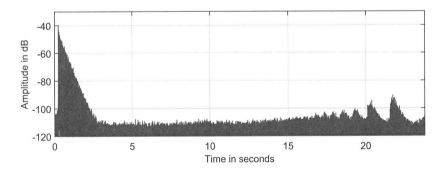

Figure 2.4.: Room impulse response measured with exponential sweep and calculated using cyclic convolution. The nonlinear components appear at the end of the impulse response.

Signal to Noise Ratio

One important aspect that describes the quality of a measurement is the ratio of signal and noise energy. In measurement theory, typically the logarithmic ratio, the so-called signal-to-noise ratio (SNR) is used:

$$\text{SNR} = 10 \cdot \log_{10} \left(\frac{\int\limits_{0}^{t_{\text{IR}}} s^2(t) \, dt}{\int\limits_{0}^{t_{\text{IR}}} n^2(t) \, dt} \right) \text{dB} \ . \tag{2.8}$$

The signal energy $s^2(t)$ and the noise energy $n^2(t)$ are integrated over the complete length of the impulse response t_{IR}. For practical applications the calculation of the SNR is not trivial, since it is not possible to record the signal without any noise components. For that reason in room acoustics often the peak signal-to-noise ratio PSNR is used:

$$\text{PSNR} = 20 \cdot \log_{10} \left(\frac{\hat{h}}{N_0} \right) \text{dB} \ . \tag{2.9}$$

It is the logarithmic ratio between the maximum signal amplitude $\hat{h}$ in the impulse response $h(t)$ and the root mean square of the stationary noise floor N_0. The PSNR can directly be read off the impulse response as the level difference between maximum and noise tail. Assuming a diffuse sound field in the room with exponential decay, PSNR can be converted to SNR and vice versa:

$$\text{SNR} = 10 \cdot \log_{10} \left(\frac{\hat{h}^2 \, \frac{T}{6 \log_e(10)} \left(1 - 10^{-\frac{6t_{\text{IR}}}{T}} \right)}{N_0^2 \cdot t_{\text{IR}}} \right) \text{dB} \tag{2.10}$$

$$= \text{PSNR} + 10 \cdot \log_{10} \left(\frac{1 - 10^{-\frac{6t_{\text{IR}}}{T}}}{\frac{t_{\text{IR}}}{T} \, 6 \cdot \log_e(10)} \right) \text{dB} \ . \tag{2.11}$$

The reverberation time T describes the decay of the room (see Section 2.2 for definition). For the same reverberation time T and a constant length of the impulse response t_{IR}, both measures differ only by a constant offset. As a rule of thumb, a ratio of impulse response length to reverberation time of 1.5, the offset can be estimated to:

$$\text{SNR} = \text{PSNR} - 13\,\text{dB} \qquad \text{for } \frac{t_{\text{IR}}}{T} \approx 1.5 \;. \qquad (2.12)$$

2.2. Acoustic Theory

Acoustic waves propagate with the speed of sound c. The relevant medium for room acoustics is air and under normal conditions the speed is approximately $c \approx 344\,\text{m/s}$. The speed of sound is dependent on the temperature Θ of the gas [21]:

$$c = \sqrt{\frac{\kappa R_{f,mol}\Theta}{M_r}} \;. \qquad (2.13)$$

The adiabatic or isentropic constant for air is $\kappa = 1.4$ and the molar mass of dry air $M_r = 0.0289644\,\text{kg/mol}$. The molar gas constant for air $R_{f,mol}$ is also dependent on temperature Θ, relative humidity φ, and static pressure p:

$$R_{f,mol} = \frac{R_{mol}}{1 - \frac{h}{100}\left(1 - \frac{R_{mol}}{M_r R_d}\right)} \qquad (2.14)$$

with the molar gas constant for air $R_{mol} = 8.31\,\text{Nm/(mol K)}$ and the gas constant of water vapor $R_d = 461\,\text{J/(kg K)}$. The molar concentration of water vapor h in percent is:

$$h = 100 \cdot \varphi \cdot \frac{p_r}{p} \cdot 10^{\left(-6.8346\left(\frac{\Theta_0}{\Theta}\right)^{1.261} + 4.6151\right)} \qquad (2.15)$$

with the reference static pressure $p_r = 101325\,\text{Pa}$.

The acoustic conditions in a room are mainly influenced by the positions and the acoustic properties of its surrounding walls. Using a geometrical acoustic approach, the sound field in a closed space can be described as a large number of particles emitted from the source position in all directions. These particles (or rays) travel along straight lines and are reflected by the walls until they reach the receiver. In an ideal diffuse sound field, the statistical mean free path length between two reflections is defined by the geometry of the room (volume V and surface S) and can be estimated to:

$$\bar{l} = \frac{4V}{S} \ .$$

(2.16)

With each reflection the particle loses a part of its energy, depending on the absorption coefficient of the wall α. This absorption causes the energy of the sound field to decay exponentially as soon as the sound source is switched off. The time that is needed for the sound field to decay to one millionth of its original value is called reverberation time. It is the most important parameter to describe the perceived acoustics of a room (c.f. Section 2.3.1) and an important measure to calculate the physical energy density of a room. Combining the mean free path length and the average absorption coefficient of the walls $\bar{\alpha}$, the theoretical reverberation time can be estimated to

$$T = \frac{-24 \cdot \log_e(10)}{c} \frac{V}{S \cdot \ln(1 - \bar{\alpha})} \ .$$

(2.17)

The so-called Eyring formula is often linearized for small absorption values to the Sabine formula:

$$T = \frac{24 \cdot \log_e(10)}{c} \frac{V}{S \cdot \bar{\alpha}} \ .$$

(2.18)

With the knowledge of the reverberation time T, the ideal room impulse response can be described as an exponential decay with maximum amplitude $\hat{h}$:

$$h(t) = \hat{h} \cdot 10^{\left(-3\frac{t}{T}\right)}$$

(2.19)

In this model approach the so far only considered propagation loss is the absorption of the walls. Formula Eq. (2.17) can be extended to cover the effect of air absorption:

$$T = \frac{-24 \cdot \log_e(10)}{c} \frac{V}{S \cdot \ln(1 - \bar{\alpha}) - 4mV} \tag{2.20}$$

where m is the damping constant of air. While this effect might be negligible for low frequencies, it becomes dominant for high frequencies. Bass et al. [22] developed formulas to calculate the damping constant for air as a function of the ambient atmospheric temperature Θ, the pressure p_a and the molar concentration of water vapor h in percent:

$$m = 2 \cdot f^2 \left[\left(184 \cdot 10^{-13} \frac{p_r}{p_a} \sqrt{\frac{\Theta}{\Theta_0}} \right) + \left(\frac{\Theta}{\Theta_0} \right)^{-5/2} \cdot \right. \tag{2.21}$$

$$\left. \left(1275 \cdot 10^{-5} \cdot \frac{e^{\left(\frac{-2239.1}{\Theta} \right)}}{f_{rO} + \frac{f^2}{f_{rO}}} + 0.1068 \cdot \frac{e^{\left(\frac{-3352}{\Theta} \right)}}{f_{rN} + \frac{f^2}{f_{rN}}} \right) \right]$$

with the reference ambient atmospheric pressure $p_r = 101.325\,\text{kPa}$ and the reference temperature in kelvin $\Theta_0 = 293.15\,\text{K}$. The two relaxing frequencies of oxygen and nitrogen are defined as:

$$f_{rO} = \frac{p_a}{p_r} \left(24 + 4.04 \cdot 10^4 \cdot h \frac{0.02 + h}{0.391 + h} \right) \tag{2.22}$$

$$f_{rN} = \frac{p_a}{p_r} \left(\frac{\Theta}{\Theta_0} \right)^{-1/2} \cdot \left(9 + 280 \cdot h \cdot e^{-4.17 \left[\left(\frac{\Theta}{\Theta_0} \right)^{-1/3} - 1 \right]} \right) \tag{2.23}$$

These calculation instructions are regulated in the international standard ISO 9613-1 [23]. The difference between Eq. (2.21) and the ISO equation is a factor of $\frac{2}{8.686}$, which results from a conversion between decadic and Naperian logarithm, and the consideration of energies instead of amplitudes.

2.3. Room Acoustic Parameters

The aim of room acoustic parameters is - as the name suggests - to describe the acoustic conditions in a room. The characterization is on one hand motivated

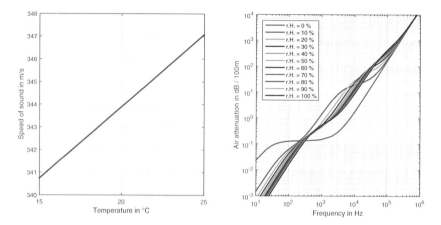

Figure 2.5.: Speed of sound in dependence of the temperature (left) and air attenuation constant m as function of frequency and for several values of relative humidity (right).

by the attempt of quantification of the subjective human perception and on the other hand the description of the physical sound field. Over the decades scientists have studied the main factors of room acoustic perception and have developed parameters to describe these influences quantitatively by using room acoustic measurement procedures. First objects of interest were especially auditoria where the acoustic condition is the key factor, such as concert halls, opera houses, theaters, and churches. Nowadays, the importance of acoustic conditions has been recognized even for smaller rooms such as conference rooms, classrooms, or sport halls. Therefore, standards were defined with appropriate ranges for room acoustic parameters (c.f. ISO 18041 [24]).

The reverberation time is used to describe the physical energy conditions in a room. Subsection, 2.3.1 gives an overview of technical measurement procedures that are based on reverberation time measurements.

Besides the measurement equipment and procedures (as described in Section 2.1), ISO 3382-1 [3] also describes definitions for a number of room acoustic parameters. These commonly accepted parameters are investigated in this thesis and are described in detail in the following section.

2.3.1. Reverberation Time

The reverberation time T has a special position among all room acoustic parameters, since it is the first known and the most important acoustic parameter. It is defined as the time that is needed for a stationary sound field to decay to one millionth of its initial energy corresponding to a level decay of 60 dB.

The old measurement principle of Sabine is still applicable according to ISO 3382 (called interrupted noise method), but it is rarely used. The random nature of the excitation noise introduces a random error to the evaluated reverberation time and is the reason for the necessity of averaging several measurements.

Schroeder [25] proposed a new method for measuring the reverberation time, where the decay curve $E(t)$ (short EDC, also known as backward integrated impulse response or Schroeder curve) is calculated from a measured room impulse response $h(t)$ (Figure 2.6 top left). The impulse response is squared to obtain the echogram (Figure 2.6 top right). The EDC is calculated by the backwards integration:

$$E(t) = \int\limits_{t}^{\infty} h^2(\tau) \, d\tau = \int\limits_{\infty}^{t} h^2(\tau) \, d(-\tau) \; . \qquad (2.24)$$

Figure 2.6 shows the result with the exponential decay of the sound energy (bottom left) or the linear decay of the level (bottom right).

$E(t)$ is often normalized to $\max(E(t)) = E(0) \overset{!}{=} 1$ to obtain as EDC that starts at $0\,\mathrm{dB}$ in logarithmic representation and has linear decay.

Schroeder showed that the average of an infinite number of interrupted random noise measurements gives the same result as one measurement of the integrated impulse response method. Based on the calculated decay curve, the decay rate is determined using linear regression.

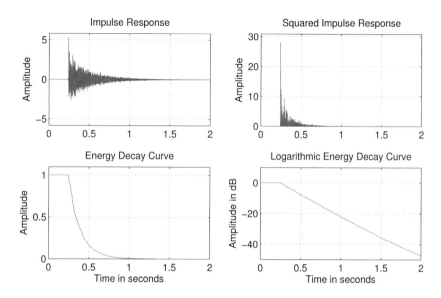

Figure 2.6.: The single processing steps to calculate the energy decay curve. The impulse response (top left) is squared (top right) and the backward integration is performed (bottom left). In the logarithmic scale (bottom right), the linear level decay can be seen.

The initial decay of the EDC is strongly influenced by the direct sound and not by the relevant reverberance. Therefore, the initial 5 dB drop in level is excluded and the linear regression is applied to the decay curve from -5 dB to -65 dB below the maximum value. The evaluated reverberation time is labeled T_{60} to indicate the evaluation range of 60 dB.

The usable decay range is limited by the level difference between signal and noise and in practice it often does not allow a determination of T_{60}. In these cases it is possible and common practice to evaluate a smaller level range, like for example 20 dB (from -5 dB to -25 dB). The decay time of 20 dB is extrapolated to 60 dB decay and the parameter is called T_{20}. Using the same method (always starting at -5 dB and evaluating the range that is specified in the index and extrapolating to 60 dB), other decay ranges (such as T_{10}, T_{15}, T_{30}, T_{40} or T_{50}) can be evaluated to choose the largest dynamic range for the given signal-to-noise ratios.

Despite the adjustment of the evaluation range, the noise contained in the measurement can have a significant influence on the calculated reverberation time. Therefore, ISO 3382 suggests a level difference of at least 10 dB between lower evaluation range and background noise level. This standard also defines two methods to compensate noise effects. These techniques and their performance are discussed in detail in Chapter 3.

Jordan [26] proposes the early decay time (EDT) with a slightly different definition of the reverberation time that also contains the direct sound. The evaluation range is from 0 dB to -10 dB and the decay time is also extrapolated to a 60 dB decay. It is shown that EDT correlates better with the subjective perception of reverberance than the reverberation time T.

According to ISO 3382 the analyzed frequency range should cover at least 250 Hz to 2 kHz, if no specific frequency bands are required. The extended frequency range for engineering and precision methods should include all bands from 125 Hz to 4 kHz.

Technical Application of Reverberation Time

In technical applications the measured reverberation time is used to derive a measure of the diffuse sound field. Therefore, the Eyring or Sabine formula is used to estimate the sound field energy (see Section 2.2). In the following sections, the most common measurement procedures are described.

Diffuse Sound Absorption The diffuse sound absorption properties of a material are determined in a reverberation chamber. This is a special measurement room equipped with sound diffusors and low absorptive walls to provide a sound field as diffuse as possible. For the determination of the absorption coefficients, the reverberation time of the empty chamber is compared to the reverberation time when the investigated sample is placed in the room. The change in the reverberation time can be lead back to the additionally inserted equivalent absorption area of the sample. The details of this measurement procedure are described in ISO 354 [27].

Sound Scattering The scattering coefficient describes the directional reflection properties of materials and surfaces. Therefore, the reverberation times in the reverberation chamber at four different conditions has to be measured. This method is standardized in ISO 17497-1 [28].

Sound Power The sound power describes the total sound energy radiated by a device. One method to determine the sound power is defined in ISO 3741 [29]. The sound pressure level caused by the device is measured in a reverberation chamber and the influence of the room is compensated by taking the reverberation time of the chamber into consideration.

Sound Insulation The determination of the insulation index of walls in buildings or building elements is described in ISO 140 [30]. Therefore, in the sending room a loud sound field is generated and the sound pressure levels in the sending and receiving rooms are measured. The room acoustic properties of the receiving room have an influence on the measured sound pressure level, but they are not related to the sound insulation. Therefore, the recorded sound pressure level has to be corrected using the measured reverberation time in the receiver room.

2.3.2. Sound Strength

Sound strength is a room acoustic parameter that characterizes the perceived loudness. It describes the stationary sound pressure level (SPL) in an auditorium with reference to the SPL of the same omnidirectional source in free field conditions and at a distance of $10\,\mathrm{m}$. The sound strength G in dB can be calculated from the impulse response:

$$G = 10 \cdot \log_{10} \left(\frac{\int\limits_0^\infty h^2(t)\,\mathrm{d}t}{\int\limits_0^\infty h_{10}^2(t)\,\mathrm{d}t} \right) \tag{2.25}$$

Here $h_{10}(t)$ is the measured impulse response of the sound source at a distance of $10\,\mathrm{m}$ under free field conditions. In order to average the inevitable directivity of the source, it is necessary to repeat the measurements every $12.5\,^\circ$ and calculate the energetic average. According to ISO 3382-1 [3], it is also possible to measure at smaller distances ($\geq 3\,\mathrm{m}$) and compensate the amplitude assuming a point source or measure the source level in a reverberation chamber. The sound strength can also be calculated directly from the energy decay curve $E(t)$:

$$G = 10 \cdot \log_{10} \left(E(t=0) \right) - L_F \tag{2.26}$$

where $L_F = 10 \cdot \log_{10} \left(\int\limits_0^\infty h_{10}^2(t)\,\mathrm{d}t \right)$ is the level of total energy in the free field reference measurement.

2.3.3. Energy Ratios

The temporal occurrence of reflections and their amplitudes have a major influence on the room acoustic perception. Early reflections that appear directly after the direct sound are not discernible as single reflections, but enhance the perceived loudness of the direct sound. The human ear integrates these first reflections and thus they help to understand speech or music and the subjective clarity is increased. Whereas, later reflections that can be distinguished as separate echoes decrease the intelligibility. The echoes and the reverberation smear the time structure and syllables become blurred, decreasing the speech comprehensibility. Various psychoacoustic studies performed listening tests and determined the limit between useful and disturbing reflections to be in the range from 50 ms to 100 ms [2, 31].

Definition D_{50}

In 1953, Thiele [32] defined the first energy ratio called definition ("Deutlichkeit"). He set the limit to divide the reflections up to 50 ms and defined the *definition* D_{50} to be the percentage energy of the first part compared to the total energy:

$$D_{50} = \frac{\int_0^{50\,\text{ms}} h^2(t)\,\text{d}t}{\int_{50\,\text{ms}}^{\infty} h^2(t)\,\text{d}t} \tag{2.27}$$

The definition is primarily used for speech presentations and is standardized in ISO 3382-1 [3]. The definition can also be directly calculated from the energy decay curve $E(t)$:

$$D_{50} = 1 - \frac{E(t = 50\,\text{ms})}{E(t = 0\,\text{ms})} \tag{2.28}$$

Clarity Index C_{80}

For musical content the temporal limit for useful reflections is higher, since the reflections are less detectible than for speech signals. Reichardt, Alim, and Schmidt [33] proposed the clarity index C_{80} with the time limit of 80 ms and on a logarithmic scale:

$$C_{80} = 10 \cdot \log_{10} \left(\frac{\int\limits_{0}^{80\,\text{ms}} h^2(t) \, dt}{\int\limits_{80\,\text{ms}}^{\infty} h^2(t) \, dt} \right) \tag{2.29}$$

The clarity index is also defined in ISO 3382-1 [3] and is used for musical presentations. The clarity index can be calculated using the energy decay curve using the formula

$$C_{80} = 10 \cdot \log_{10} \left(\frac{E(t = 0\,\text{ms})}{E(t = 80\,\text{ms})} - 1 \right) \tag{2.30}$$

A rough assessment of the clarity index can be made by assuming an ideal exponential sound field, which is only described by the reverberation time (c.f. Eq. (2.19)) and therefore the clarity index can be calculated as:

$$C_{80} = 10 \cdot \log_{10} \left(10^{\frac{6 \cdot 80\,\text{ms}}{T}} - 1 \right) \tag{2.31}$$

Estimations of other energy parameters like D_{50} can be derived in a similar way. However, this estimation only provides a rough estimate of the whole room, since for this theory the reverberation time is the same as in the room. The direct sound and the early reflections that are the motivation for defining the room acoustic energy parameters are not included in the ideal diffuse field theory.

2.4. Assessment of Error Size

The assessment of the tolerable error size for room acoustic parameters depends strongly on the application. The decisive measure for room acoustic parameters to describe the subjective perception is the just noticeable difference (JND) of the corresponding parameter. The JND describes the amount of change in the room acoustic parameter that is necessary to notice a perceptive difference. JNDs (also known as difference limens) are determined by listening tests and are defined as the difference where half of the test persons recognize the diversity correctly.

Seraphim [34] investigated the difference limens of reverberation time in 1958. He found JNDs of around 5% in the typical range of reverberation times from 0.5 s to 2 s. Since for listening tests synthetic stimuli with exponential decays are used, all reverberation times (EDT, T_{10}, T_{20}, etc.) are equal and the derived results apply to all reverberation times. Cox, Davies, and Lam [35] analyzed the JND of the clarity index and found thresholds of about 1 dB and only a small dependence on the motif. Further studies (for example, by Bradley, Reich, and Norcross [36] or Ahearn et al. [37]) confirmed the findings. These just noticeable differences are also reported in ISO 3382 and are generally accepted. For the assessment of errors in room acoustic measurements, these form the basis to define the maximum tolerable error. For measured room acoustic parameters that deviate in the order of JNDs from the correct value, even the rather insensitive human perception is able to sense a difference and these measurements are unusable. To allow a description of the perceived room acoustic or to compare several measurements on a perceptual basis, it is necessary that the error is at least less than half the JND. For single interfering influences that fall below one tenth of the JND, it can be assumed that the influence is negligible.

The reverberation time is also used in numerous technical applications (c.f. Subsection 2.3.1) and a measurement error in the reverberation time propagates directly to the calculated parameter. Thereby the sensitivity of the result on the input error depends on the corresponding calculation. Depending on the determined parameter and on the intended purpose of the results, the tolerable error of the reverberation time can be significantly lower than for the formerly mentioned perceptual purpose.

The most sensitive applications for errors are methods where the detailed time structure of the impulse response is analyzed directly without preprocessing steps such as the backward integration or linear regression over a wider range.[2] Techniques that superposition several single measurements to calculate the final result are very sensitive to small changes. An example is the technique that uses a large number of measurements with different source directivity to maintain a spherical harmonics representation of the source [38]. With the spherical harmonics coefficient any source directivity can be synthesized afterwards.[3] Even extremely small changes of the acoustic system invalidate the time-invariance assumption that is crucial for the superpositioning approach. Changes of room acoustic parameters can be considered as indicators for time variant systems and point out critical measurement sessions.

---

[2] Of course, the same applies to methods that operate in frequency domain.

[3] The possible directivities depend on the order of the available spherical harmonics domain.

3

Noise Compensation

This chapter analyzes the effects of stationary noise as external influence on the evaluation of room acoustic parameters. Stationary noise is inevitable in every measurement and has a variety of sources (see Section 2.1.2). There are numerous methods to compensate the effects of stationary noise. This study investigates five established and commonly used noise compensation methods and compares their performances. Three of these techniques are also defined in ISO 3382-1 [3]. The findings of this chapter have been partly published in Guski et al. [39].

In 2004, Katz carried out a round robin for room acoustic analysis software [7]. All participants (19 institutions with 25 different software packages) used the same measured room impulse response and reported the evaluated parameters. The differences between the software packages remained within the range or exceeded the subjective difference limen of the corresponding parameter. Katz pointed out that the variations occur mostly due to the noise in the impulse response. Another major noise-related problem was the lack of indication of noise effects that are too high. Reverberation time results can be obtained with most software packages, even if the signal-to-noise ratio is insufficient.

The effects of noise and techniques to suppress it were subject to several studies [5, 40, 41, 42]. Hak, Wenmaekers, and Luxemburg [43] performed the first systematic analysis of one noise compensation technique depending on the noise level. Random white noise with different levels was added to a real measured impulse response to simulate different signal-to-noise ratios. It is also possible to calculate the reverberation time with nonlinear regression methods [44, 45, 46], but these methods are rarely used and therefore not investigated in this thesis.

3.1. Noise Compensation Methods

All five investigated noise compensation methods use the impulse response as input, apply the compensation, and provide the noise compensated energy decay curve $E(t)$ (EDC). The reverberation time is determined by performing a linear regression of $E(t)$ (see Section 2.3.1). Other room acoustic parameters (such as the definition D_{50}, clarity index C_{80} or the sound strength G) can be calculated directly from the noise compensated decay curve (see Section 2.3), so that the calculation of all parameters benefits from the same noise compensated EDC.

In the following subsections, the investigated methods will be described and illustrated using a sample impulse response. The impulse response represents a reverberation time of 2 s and a peak signal-to-noise ratio of 40 dB. The model impulse response represents any measurement in the statistical region of the room transfer function where neither modal effects of the room nor effects of too narrow filter bands are present. The corresponding EDC is also shown to illustrate the error that occurs due to the noise.

Some methods require additional parameters of the impulse response (such as noise level, intersection time, and late reverberation time). These are estimated using the iterative algorithm proposed by Lundeby et al. [5]. This algorithm has been proven to provide reliable results for real measured room impulse responses. The complete evaluation is fully automated and realized in Matlab. All described noise compensation methods are part of the ITA-Toolbox, an open source toolbox for Matlab [11]. Method A is the default method described in ISO 3382-1 [3]. Methods B and C are optional extensions also defined in the standard. Methods D and E use a noise suppression technique that is not described in ISO 3382-1.

3.1.1. Method A: Full Impulse Response

The full impulse response $h(t)$ is considered and the integration limits are determined by the length of the recorded impulse response t_{IR}:

$$E(t) = \int\limits_{t}^{t_{\mathrm{IR}}} h^2(\tau)\,\mathrm{d}\tau \ . \tag{3.1}$$

Technically speaking, this is not a noise compensation method as no noise compensation technique is used. The noise contained in the impulse response causes a large overestimation of the EDC (Figure 3.1). The error size increases with the length of the impulse response t_{IR}.

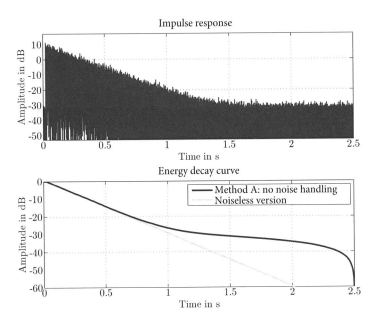

Figure 3.1.: Example of an impulse response and corresponding energy decay curve when using the full impulse response for the backward integration (Method A). The last part of the energy decay curve is overestimated.

3.1.2. Method B: Truncation at Intersection Time

A commonly used noise compensation method is the truncation of the impulse response at the intersection time t_i [3, 5, 41]:

$$E(t) = \int\limits_{t}^{t_i} h^2(\tau) \, \mathrm{d}\tau \; . \tag{3.2}$$

The intersection time is the time where the exponential decay of the impulse response intersects with the constant noise floor. The noise error in EDC is reduced drastically (Fig. 3.2), but the truncation introduces a further error. The energy decay curve is approaching minus infinity because of the missing signal energy from the truncation time to infinity. The unlimited dynamic range of the resulting EDC always allows an evaluation of the EDC, even if the signal-to-noise ratio is insufficient for a certain reverberation time.

3.1.3. Method C: Truncation and Correction

Lundeby et al. [5] proposed a correction term to prevent the truncation error. The missing signal energy from truncation time to infinity (triangle in Figure 3.3) C_{comp} is estimated and added to the truncated integral:

$$E(t) = \int\limits_{t}^{t_i} h^2(\tau) \, \mathrm{d}\tau + C_{\mathrm{comp}} \; . \tag{3.3}$$

ISO 3382 recommends calculating C_{comp} by assuming an exponential decay. The decay rate should be the same as given by the squared impulse response between t_1 and the intersection time. The time denoted by t_1 corresponds to a level, which is 10 dB above the level at the intersection time. This part of the impulse response, however, is already influenced by noise. At the intersection, time signal and noise energy are equal by definition. Hence, Lundeby et al. suggested to

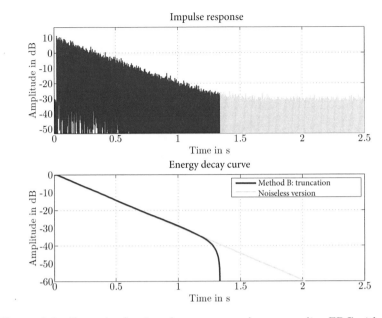

Figure 3.2.: Example of an impulse response and corresponding EDC with truncation at the intersection time (Method B). The influence of noise is reduced significantly. The energy decay curve is underestimated.

leave a safety margin of 5 - 10 dB above the level corresponding to the intersection time to reduce the influence of noise. In this study a safety margin of 10 dB is used, if not further specified.

The resulting EDC shows no truncation error any more (see Figure 3.3). The dynamic range of the EDC is limited according to the signal to noise ratio. The error caused by the noise is reduced significantly, even though a slight overestimation due to noise before truncation time is still present.

39

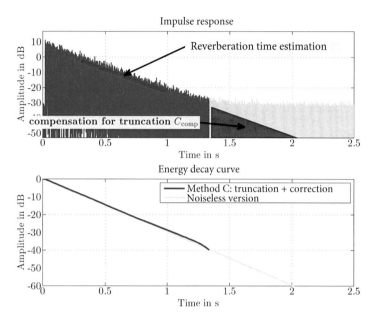

Figure 3.3.: Example of an impulse response and corresponding EDC for truncation and correction term for the truncation (Method C). The influence of noise is reduced; just a small overestimation can be observed. The truncation error is eliminated and the EDC is limited to a reasonable range.

3.1.4. Method D: Subtraction of Noise

Chu [42] proposed the "subtraction of noise"-method. The noise level N_{est} is estimated and subtracted from the impulse response before backward integration:

$$E(t) = \int\limits_{t}^{t_{\text{IR}}} \left(h^2(\tau) - N_{\text{est}}^2 \right) \, \mathrm{d}\tau \ . \tag{3.4}$$

The noise component of the original measured impulse response is squared and therefore always positive. The error sums up, due to the integration and the EDC is therefore overestimated with this method. By subtracting the estimated noise level, the distribution of the new noise component can be considered as zero mean. The noise error is reduced due to temporal averaging during the integration. This technique works well for the first part of the impulse response where the signal energy is dominant. For the later part, after the intersection time, this technique fails because the EDC is approaching minus infinity and is not monotonically decreasing any more (see Figure 3.4). Similar to Method B, the problem of an unlimited dynamic range of EDC occurs.

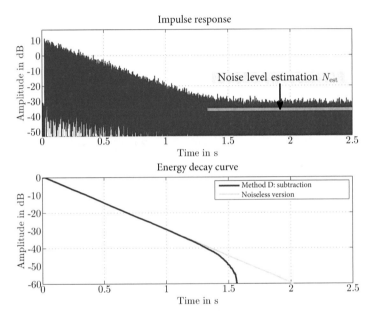

Figure 3.4.: Example of an impulse response and corresponding EDC with subtraction of estimated noise level (Method D). The influence of noise on the EDC is minimized, but the dynamic range of EDC is unlimited.

3.1.5. Method E: Truncation, Correction, and Subtraction

The fifth method is a combination of all techniques mentioned above: The estimated noise level is subtracted before backward integration, the impulse response is truncated at the intersection time, and the correction for the truncation is applied:

$$E(t) = \int\limits_{t}^{t_i} \left(h^2(\tau) - N_{\text{est}}^2 \right) \, \mathrm{d}\tau + C_{\text{comp}} \ . \tag{3.5}$$

The noise influence is minimized by subtracting the noise level. The truncation of the impulse response suppresses errors in the later part of the room impulse response resulting from noise subtraction and a correction for the truncation ensures a reasonable dynamic range (Figure 3.5).

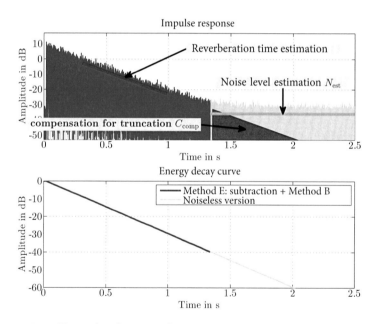

Figure 3.5.: Example of an impulse response and corresponding EDC with truncation at intersection time, correction for truncation, and subtraction of noise level (Method E). The influence of noise is minimized and the dynamic range of EDC is limited.

3.2. Evaluation Approaches

The analysis of the different noise compensation techniques is carried out using two evaluation approaches that are described in the following two sections.

3.2.1. Model Approach

The model approach is a parametric description of an ideal room impulse response with added noise. Only the envelope of the impulse response is modeled without any temporal fine structure. The signal decay is assumed exponential and is described by the reverberation time T. The noise is stationary and its level is defined using the peak signal-to-noise ratio (PSNR). The maximum of the signal part is compared with the mean level of the noise. The third parameter is the length of the measured impulse response (t_{IR}), since it has an impact on the results obtained using one of the methods. The model room impulse response of Eq. (2.19) can be extended to:

$$h(t) = 10^{\frac{-3t}{T}} + 10^{\frac{-\text{PSNR}}{20}} \quad . \tag{3.6}$$

The linear regression is performed numerically using a sufficiently high sampling rate to exclude influences that occur due to sampling [47]. The difference between the envelope and the mean value of the background noise is assumed approximately 11 dB, supposing a Gaussian distribution.

3.2.2. Measurement Approach

Most measurement approaches described in the relevant literature only investigate a few examples of different noise levels. This study tries to present a systematic approach, as in the study published by Hak, Wenmaekers, and Luxemburg [43], rather than some exemplative results that only show that there is a difference. In contrast to Hak's study where additive white noise is used, the

43

different signal-to-noise ratios are obtained by performing real measurements and using real background noise. This procedure also allows a reliable statistical analysis of the variances for real measurements.

These long-term measurements have been conducted in the General Assembly Hall of RWTH Aachen University (for detailed information see Appendix A). The receiver measurement chain side consisted of one $\frac{1}{2}$-inch condenser microphone (B&K Type 4190) and 15 Sennheiser KE4 microphones. The remaining hardware has been equivalent to the description in Section 2.1.2. The complete input and output measurement chain has been calibrated to make it possible to measure absolute levels. For the excitation, an exponential sine sweep with a frequency range from 20 Hz to 14 kHz and a length of six seconds has been used. In the post-processing the impulse responses have been truncated at four seconds to guarantee a robust noise detection and allow neglecting the influences of nonlinear components at the end. These measurements have been performed using the measurement application of the ITA-Toolbox [11].

For each measurement the amplification of the excitation signal has been changed to obtain different peak signal-to-noise ratios. A set of seven different amplifications was repeated 72 times to allow a statistical statement of the resulting parameters.

The PSNR is calculated separately for each measured impulse response to account for changing noise levels during the measurement session and the measurements have been clustered into groups according to their PSNR values. For each group the mean value and the standard deviation of the evaluated room acoustic parameter are calculated. The maximum difference in PSNR in one group is 1 dB. Due to this small range in one group, a high resolution of the PSNR dependency is obtained and smearing effects are avoided. In the available measurement time, 504 broadband impulse responses have been obtained. They have the same signal content, while the noise part changes for every measurement under realistic conditions (temporal distribution and frequency spectrum). The 16 microphones have been spread over the listener area and have been used simultaneously.

The most important prerequisite for the success of this approach is the acoustic stability of the room. If the room changes while the measurement is carried out, it is not possible to distinguish these changes from the effects of noise. To ensure this stability, those measurements have been performed automatically without any people present. In addition, the first 90 minutes after the last person had left the room have been discarded to make sure that the room conditions had settled. Temperature and humidity have been monitored during the measurement and can be considered constant within a maximum difference of 0.6 °C and 6 % relative humidity. The different output levels appear fragmented and spread over the complete measurement time on purpose. This way, long-term changes such as the temperature drift have no systematic effect on certain PSNR conditions but appear randomly spread over all measurements.

3.3. Comparison of Noise Compensation Methods

The mean value and the standard deviation of the evaluated reverberation time T_{20} for the described measurement approach are plotted in Figure 3.6 as a function of the peak signal-to-noise ratio. Most compensation methods show a clear dependence of the evaluated reverberation time on the PSNR. For very high PSNRs, however, T_{20} converges towards a constant value. This implies that the noise does not affect the evaluation for sufficiently high PSNRs. These results are in accordance with the conclusions of [43]. Furthermore, all noise compensation methods show the same results for high PSNR values. This also demonstrates that the compensation method has no influence on the evaluation for sufficient high PSNRs. The fact that the results are not affected by the output level for high PSNRs also shows that nonlinearities of the loudspeaker (that increase with output level) have no influence on the parameter. The resulting parameters for the highest PSNR are considered as the true value and are used as a reference (separately for each noise compensation method) to show the relative error of the room acoustic parameters in the following analysis. Other room acoustic parameters show the same dependencies and are therefore referred to their best estimate.

3.3.1. Reverberation Time

First, the reverberation time T_{40} is investigated. Although T_{40} is not very often used in practice, it is very useful to compare the results of the model and

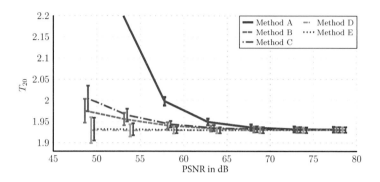

Figure 3.6.: Evaluated reverberation time T_{20} in dependence of the peak signal-to-noise ratio for all five analyzed methods. The error bars indicate the standard deviation.

measurement evaluation approaches. This parameter covers a wide range of PSNR values from insufficient to very good signal-to-noise ratios and thus allows observing all occurring effects. ISO 3382 recommends a decay range of 55 dB for the parameter T_{40}, which is equal to a PSNR of 55 dB in diffuse sound fields [3].

Figure 3.7 shows the relative error of T_{40} for all five methods. Using no noise compensation (Method A) results in a huge overestimation of the reverberation time. This is in accordance with the effect seen in the EDC (see Section 3.1). The truncation of the impulse response at the intersection time (Method B) reduces the noise effect considerably for high and medium PSNRs. For low PSNRs the reverberation time is strongly underestimated. This effect is caused by the underestimation in the energy decay curve. Applying the correction for the truncation (Method C) prevents this effect. Method C automatically yields no results for an insufficient dynamic range of the EDC. The slightly bigger errors for medium PSNRs compared to Method B are caused by the absence of one of the two opposing effects (overestimation caused by noise and now missing underestimation caused by truncation). The subtraction of noise technique (Method D) gives perfect results for the model approach. However, this is caused by the simplicity of the model and the results cannot be interpreted. The measurement approach shows that there is nearly no systematic error for medium and low PSNRs. For low PSNRs the unlimited range of EDC results in an underestimation, similar to Method C. Method E (truncation, correction, and subtraction of noise) yields the best results of the measurement approach. There is nearly no systematic error for mid and high PSNRs and for insufficient PSNRs the algorithm automatically provides no results. The results of the model approach can again only partly be interpreted due to the simplicity of the model.

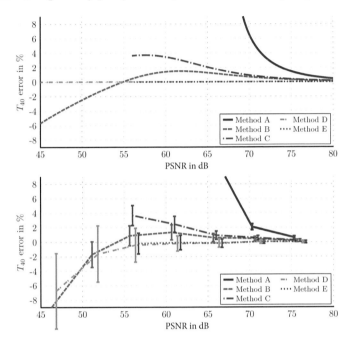

Figure 3.7.: Relative error of reverberation time T_{40} for model (top) and measurement approach (bottom). The measurement approach shows the 250 Hz octave band.

The results that are obtained using the model and measurement approaches are consistent with each other and thus the measurements confirm the validity of the model. An exception is Method D, where the simple constant noise approach of the model results in perfect output parameters.

In the next step, the reverberation time T_{20} is examined, since it is of greater relevance in practice. In Figure 3.8 it can be seen that the measurement approach only provides data for high to medium PSNR range. The model results have to be investigated to analyze the lower and therefore more interesting PSNR range. Compared to T_{40} the results are shifted on the PSNR axis, which is a result of the different dynamic range of T_{20}. The smaller evaluation range and the resulting bigger percentage of the noise-influenced part is the reason for the greater relative errors.

A relative error of 5% is used to determine the required minimal PSNRs for each method, since 5% is also the commonly accepted just noticeable difference for the reverberation time. The systematic error for Method A (no noise compensation) exceeds the 5% limit at 55 dB. This is 20 dB higher than the PSNR of 35 dB recommended in ISO 3382. Since the error is dependent on the total length of the impulse response, these limits are only valid for this example, where the impulse response length $t_{IR} = 4$ s and the reverberation time $T = 2$ s. For the truncation technique (Method B), the limit is in accordance with the ISO recommendation. When using truncation and correction (Method C), the 5% limit is at about 45 dB. However, the systematic error of this method will never exceed 8%, whereas for the previous methods the error can increase significantly. No systematic error is predicted for Method E (truncation, correction, and subtraction of noise) by the model approach. The measurement approach confirms these results for PSNRs where measurements are available. The systematic error will be clearly below 5% and the results will be discarded for insufficient PSNRs, similarly to the T_{40} results.

The mean errors in the reverberation time are systematic while the errors described by Hak in his study are random and have a zero mean. These differences may be caused by the artificial character of the additive noise or by the differences in the implementation of the calculation algorithms. Another reason might be that in the investigation of Hak many impulse responses from different room and reverberation times were used.

3.3.2. Clarity Index

Deviations due to noise appear at the end of the EDC and move to the early parts for increasing noise levels. Because the clarity index C_{80} is based on only one

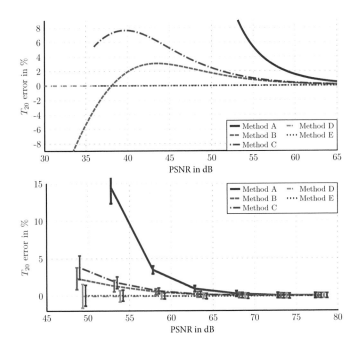

Figure 3.8.: Relative error of reverberation time T_{20} for model (top) and measurement approach (bottom). The measurement results show only high PSNRs due to the chosen output amplitudes during the measurements. (The ranges of the x-axis are not equal to better illustrate the desired effects.)

value of the very early normalized EDC, it is assumed that C_{80} is less sensitive to noise.

Figure 3.9 shows the evaluation of the results. Method A (no noise treatment) again shows the greatest sensitivity to noise. C_{80} is underestimated and the results fall below the limit of just noticeable difference of 1 dB at a peak signal-to-noise ratio of approximately 32 dB. The minimal required PSNRs decrease by approximately 10 dB to 22 dB for Method B (truncation) and to 20 dB for Method C (truncation and correction). The advantage of the automatic limitation of the results for insufficient peak signal-to-noise ratios for Methods C and E do not work for C_{80}. This means that for the clarity index (and analog for the definition D_{50}) the PSNR has always to be evaluated to estimate the noise error. For inadequate PSNRs, the C_{80} results have to be discarded. Methods B, C, D and E perform similarly and do not differ significantly. Again, the model and measurement evaluation approaches are very similar.

3.3.3. Sound Strength

The sound strength G represents basically the total energy contained in the measured room impulse response. The correct value is the pure signal energy, but since every real measurement also contains noise, the evaluated energy is the sum of signal and noise energy. Since in typical room impulse responses the signal energy is significantly higher than the noise energy, the error in G is small. Figure 3.10 shows that Method A (no noise compensation) results in the largest error. The JND for G of 1 dB is exceeded at around PSNR = 17 dB. The other methods are very robust and the error stays below 0.1 dB for PSNR values above 12 dB.

The data from the measurement approach covers only high PSNR values and does therefore not provide any new information.

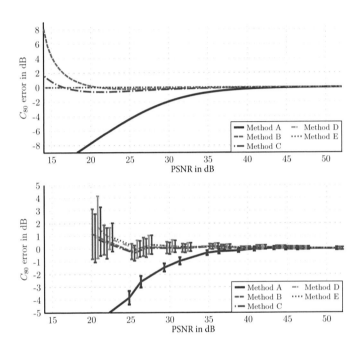

Figure 3.9.: Error of clarity index C_{80} for model (top) and measurement approach (bottom). Using no noise compensation (Method A) causes the largest systematic errors. All noise compensation methods (Methods B - E) reduce the noise influence and do not differ significantly.

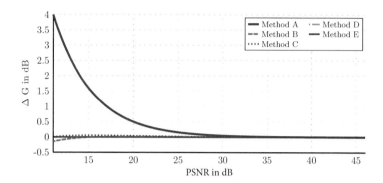

Figure 3.10.: Error of sound strength G for model. Using no noise compensation (Method A) causes the largest systematic errors. All noise compensation methods (Methods B - E) reduce the noise influence significantly.

3.4. Summary of Noise Compensation Methods

In the previous chapters single measurement results were compared to the model approach to confirm the validity. In this chapter the acquired findings are summarized to allow a general statement of the error size and the influence factors. Therefore, the systematic and the random errors caused by stationary noise are analyzed separately.

Systematic Errors

The magnitude of systematic errors caused by the influence of stationary noise is larger compared to the random error in the majority of cases. Therefore, it is necessary to document the functional relationship between error and PSNR values to allow straightforward error estimation. To describe the quantitative magnitude of the systematic error, the results of the model approach are utilized. The advantage is that the model can be evaluated for arbitrary PSNR values, whereas the measurement results are only available for discrete PSNR values that occurred during the measurements.

As alternative to the rather complex evaluation of the model approach, the following illustrative model evaluations are presented in tables to allow an easy error estimation. Table 3.1 summarizes the systematic relative error of the reverberation time T_{20} for the given example of a reverberation time of $T = 2\,\text{s}$ and the five analyzed noise compensation methods as function of the peak signal-to-noise ratio. For situations where the PSNR values are too low and the methods provide no results, the systematic errors are marked as NaN (not a number). In a similar way, Table 3.2 provides the errors for the clarity index C_{80}. Further parameters are described in Appendix B.1.

PSNR dB	Method A %	Method B %	Method C %	Method D %	Method E %
20	300.809	-66.363	NaN	-0.000	NaN
25	494.417	-40.909	NaN	-0.000	NaN
30	795.750	-19.620	NaN	-0.000	NaN
35	793.036	-5.315	NaN	-0.000	NaN
40	728.759	1.925	7.652	-0.000	0.004
45	482.815	2.920	5.201	-0.000	0.000
50	29.536	1.720	2.432	-0.000	0.000
55	5.344	0.784	0.999	-0.000	0.000
60	1.518	0.322	0.388	-0.000	0.000
65	0.466	0.125	0.146	-0.000	0.000

Table 3.1.: Systematic error of reverberation time T_{20} in percent as function of the peak signal-to-noise ratio (PSNR) and for the five investigated noise compensation methods.

3.4.1. Random Errors

The measurement approach is used to evaluate the random fluctuations of the room acoustic parameters, since the model approach only supplies information about systematic errors. Therefore, all relevant data of six octave band frequencies and several microphone positions is combined and analyzed together. Figure 3.11 shows the empirically determined relative standard deviations of the reverberation time T_{20} in dependence of the peak signal-to-noise ratio for the compensation

PSNR dB	Method A dB	Method B dB	Method C dB	Method D dB	Method E dB
15	-11.14	4.30	0.83	0.00	0.00
20	-7.74	0.19	-0.59	0.00	0.00
25	-4.36	-0.26	-0.52	0.00	0.00
30	-1.95	-0.19	-0.28	0.00	0.00
35	-0.72	-0.10	-0.12	0.00	0.00
40	-0.24	-0.04	-0.05	0.00	0.00
45	-0.08	-0.02	-0.02	0.00	0.00
50	-0.03	-0.01	-0.01	0.00	0.00
55	-0.01	0.00	0.00	0.00	0.00
60	0.00	0.00	0.00	0.00	0.00

Table 3.2.: Systematic error of clarity index C_{80} in dB as function of the peak signal-to-noise ratio (PSNR) and for the five investigated noise compensation methods.

method with truncation and correction (Method C). All results show a similar behavior, although the temporal time structure and the reverberation time of the single results differ quite strongly because of the various microphone positions and frequency bands. The 95th percentile is shown as black line to indicate the upper limit of the relative standard deviation that is only exceeded by 5% of the measured data. Table 3.3 summarizes the in the same way evaluated estimations of the random error for all five noise compensation methods. With exception of Method A, all other methods show nearly similar behavior for the random fluctuations.

It can be seen that standard deviations decrease with increasing PSNR level up to PSNR $\approx$ 70 dB. Above that point, the standard deviation approaches a constant value of around 0.6%. The relative variances fall below the limit of 5% for PSNR values $\geq$ 45 dB.

To obtain an analytical relationship between PSNR values and the relative standard deviation, a function has been designed to fit the empirically evaluated

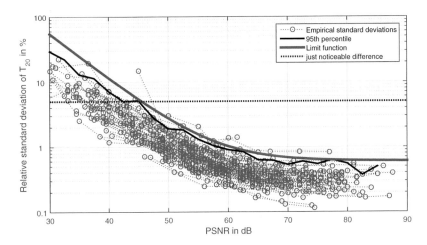

Figure 3.11.: Evaluated empirical standard deviation of T_{20} as function of PSNR values. This figure shows noise compensation Method C for several microphone positions and frequency bands. The black line marks the 95th percentile and the red line the analytic error estimation.

data from the measurements:

$$\frac{\sigma(T_{20})}{T_{20}} = \left(\frac{3 \cdot 10^5}{\text{PSNR}} \right)^{0.7} + 0.6 \;, \tag{3.7}$$

where the PSNR is inserted as an energetic ratio. This analytic function describes the upper limit of the estimated standard deviation as can be seen in Figure 3.11.

The empirically evaluated random variances of further room acoustic parameters can be found in Appendix B.2

3.5. Further Influences

In the following section, further details of the measurement or the implementation are investigated. The influences are determined based on the developed model

PSNR dB	Method A %	Method B %	Method C %	Method D %	Method E %
30	10.655	61.261	29.226	87.650	29.496
35	8.850	12.630	15.844	14.002	12.962
40	8.853	7.150	7.353	7.275	7.076
45	46.469	5.271	4.875	7.915	5.142
50	20.987	1.906	2.011	2.271	1.898
55	13.707	1.287	1.276	1.354	1.313
60	1.567	0.928	0.933	0.947	0.905
65	0.625	0.623	0.622	0.625	0.625
70	0.531	0.530	0.530	0.530	0.530
75	0.542	0.542	0.542	0.541	0.541
80	0.547	0.547	0.547	0.548	0.548
85	0.495	0.494	0.494	0.494	0.494

Table 3.3.: Relative random error of the reverberation time T_{20} as function of the PSNR value for the five investigated noise compensation methods. Values are defined empirically by determining the 95th percentile of measurement data.

and the measurements.

3.5.1. Implementation of Correction for Truncation

In Section 3.1, two different implementations to calculate the truncation correction (used for Method C and E) are discussed. The late reverberation time used to determine the correction term can be calculated without (ISO 3382-1 [3] correction) or with a safety margin of $10\,\mathrm{dB}$ above noise level (Lundeby et al. [5] correction). For the estimation of the other parameters required for the correction term (noise level and intersection time), the safety margin is always taken into account, as proposed by Lundeby, because without the safety margin the iteration algorithm does not work properly. Both evaluation approaches are used to highlight the difference in performance for both corrections. The model approach shows that the systematic error is larger if the ISO correction is used (Figure 3.12): maximum error $\approx 7\,\%$ for ISO correction and $\approx 5\,\%$ for the Lundeby correction. The same differences between the two corrections can be observed if the measurement approach is applied. The standard deviation for the ISO correction is slightly bigger than for the Lundeby correction, which is caused by the bigger influence of the noise as no safety margin is applied. The deviation between model and measurement approaches is small and does not exceed the standard deviation of the measurements.

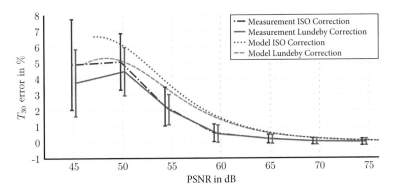

Figure 3.12.: Relative error of reverberation time T_{30} for two different implementations of the truncation correction term.

3.5.2. Excitation Signal

In this section, the effect of the utilized excitation signal on the noise sensibility is analyzed. The two most commonly used excitation signals are discussed: sine sweeps and maximum length sequences.

Figure 3.13 shows the absolute errors for the reverberation time T_{40} for MLS and sweep measurements. The frequency band and the position of the microphone are equal for both excitation signals. The deviation in PSNR values between both signals is caused by differences in the frequency spectra of the signals. Analog to the sweep measurements, the amplitude of the MLS was increased in steps of 5 dB. For lower PSNRs this leads to an increase of PSNR by the same amount. In case of higher PSNRs, however, the increase is less than 5 dB. The noise detection algorithm identifies the nonlinear spikes as noise floor, leading to an increasing noise floor for higher output amplitudes. The evaluated reverberation time does not change for higher output levels. This indicates that the nonlinear

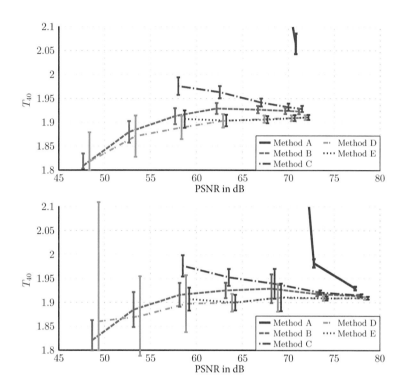

Figure 3.13.: Comparison of evaluated reverberation time T_{40} for MLS (top) and sweep (bottom) excitation. For both results the microphone position is equal and the 1 kHz octave band is shown.

components in the impulse response are still small compared to the linear signal part.

The systematic errors depending on the PSNR are similar for sweeps and MLS. The random deviations between both excitation signals hardly ever exceed 1-2 %. One reason for the similarity of the results is the control of the measurement conditions. The lack of impulsive noise in the measurement and only small nonlinear components (due to the high-quality loudspeaker) ensure that situations where the excitations display a different behavior do not occur.

3.6. Conclusion

The performances of the five noise compensation methods differ significantly. It is shown that most methods lead to systematic errors. However, these errors can be predicted. Nevertheless, it must be mentioned that the performance of the three methods allowed by ISO 3382 also deviate significantly. Using no noise compensation (Method A) results in a large overestimation of the reverberation time, depending on the total length of the impulse response. Using a noise compensation method reduces the error significantly. The ISO compliant method to truncate the impulse response at the intersection time and correct for the truncation (Method C) has the advantage that results for reverberation times are discarded automatically if the peak signal to noise ratio is insufficient. The systematic error for T_{20} is always $< 8\,\%$. Method E (subtraction of noise, truncation, and correction) showed the smallest sensitivity to noise. The systematic error is negligible and for insufficient PSNRs the results are discarded automatically. However, due to the subtraction technique this method is not compliant with the ISO standard. Table 3.4 provides an overview of all methods.

Method	Treatment	Noise sensitivity	ISO compliance	Limitation of results
Method A	none	high	yes	no
Method B	truncation	medium	yes	no
Method C	truncation and compensation	medium	yes	yes
Method D	subtraction	low	no	no
Method E	truncation, compensation, and subtraction	low	no	yes

Table 3.4.: Overview of the investigated noise compensation methods and their properties.

In general, the clarity index C_{80} is less sensitive to noise. Again, Method A (no noise treatment) shows the greatest noise sensitivity. The different noise compensation techniques (Methods B to E) are quite similar. Definition D_{50} and sound strength G are even less affected by noise.

Theoretically, the knowledge of the systematic errors allows a manipulation of the measured reverberation time by choosing a suitable noise compensation method and adjusting the measurement parameters (impulse response length or output amplification). To avoid unclear or ambiguous manipulations and to allow an estimation of the noise-induced error, the following specifications should be included or modified in the ISO 3382 standard:

- The application of a noise compensation technique should be mandatory or recommended. In the latest version of the standard, noise compensation is optional.

- The noise subtraction technique proposed by Chu should be included as a possible noise compensation technique.

- The peak signal-to-noise ratio and the applied noise compensation method should be included in the measurement report, so that an estimation of the systematic error is possible afterwards.

- The proposal on how to calculate the correction of truncation should include the safety margin above noise floor as proposed by Lundeby et al. to reduce the error.

4

Impulsive Noise Detection

The investigations in the previous chapter assume stationary ambient noise. However, this assumption is not always fulfilled for real measurements. Therefore, this chapter investigates the effects and the detection of impulsive noise during a measurement. The content of this chapter is based on the publication of the author [48].

Impulsive noise during a measurement is often caused by sources such as creaking wooden floors or beams, footsteps of people in the room, or coughing. In MLS or random noise measurements, impulsive noise is transformed into a relatively low-level random noise component in the measured response, and it has thus only small influences on further processing steps. Nevertheless, this noise is spread over all frequencies and the full period of the RIR. In contrast, for sweep signals the transient (impulsive) disruption results in a deterministic artifact in the measured impulse responses that distorts evaluation and auralization [15, 49]. Up to now, there exists no reliable method for identification and evaluation of impulsive noise in sweep measurements.

In this chapter, the automatic detection of impulsive noise in sweep measurements is described. Farina [50] showed that it is possible to properly reduce the effects of impulsive noise only by individual treatment of every measurement while automatic methods give insufficient results. A robust automatic compensation of the disruption, however, could not be found yet and will be subject to further research. The presented detection technique can be used for any excitation signal, but, due to the high sensitivity of sweeps to impulsive noise, it is especially important for sweep measurements. This technique can be used to analyze the measured response automatically, immediately after the affected measurement and to repeat the measurement period, if necessary.

All simulations, measurements, room acoustic evaluation, and data processing are done with the ITA-Toolbox [11]. The ITA-Toolbox also includes a script (ita_tutorial_impulsive_noise.m) to reproduce all results of the following chapter.

4.1. Theory

For illustration purposes in this section, a simulated impulse response is used. A rectangular room with a reverberation time of 1.5 seconds is simulated with a modal superposition approach for a specified source and receiver position [51, 52]. Figure 4.1 (top) shows the RIR using a logarithmic scale (also known as energy room-impulse response or echogram).

There are two noise components in the model: Stationary noise is unavoidable for every measurement and therefore always included in the simulation, while impulsive noise only occurs occasionally and has to be detected (see Figure 4.1 bottom).

The impulsive noise is modeled as a Dirac pulse and the influence of the room is considered by simulating the impulse response from a different source position, far away from the actual loudspeaker position. Ambient and electro-acoustic equipment noise is modeled as Gaussian white noise. This approach allows controlling the conditions and observing the effects in every processing step for every signal component (excitation signal and impulsive noise) separately. The excitation signal is an exponential sweep with a constant envelope described in Eq. (2.7). The recorded signal $r_m(t)$ at a microphone position in a room is a combination of the occurring noise $n(t)$ and the excitation signal modified by room $h(t)$:

$$r_m(t) = s(t) * h(t) + n(t) .$$

$$(4.1)$$

To obtain the desired impulse response of the system, the recorded signal has

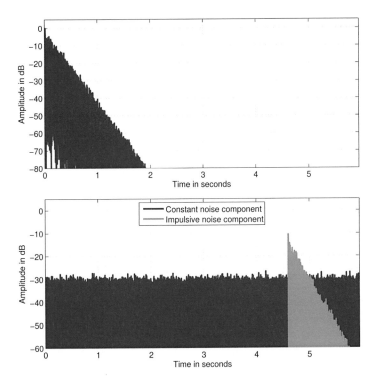

Figure 4.1.: Top: Simulated energy RIR without any noise component. Bottom: Two noise components. The stationary noise is unavoidable (blue) and the impulsive noise component (green) has to be discovered.

to be convolved with a deconvolution signal $c(t)$ (also known as compensation signal). The deconvolution signal is the one with the inverted complex spectrum of the excitation signal. For the commonly used exponential sweep with constant envelope (Eq. (2.7)), the deconvolution signal is a time-reversed exponential sweep with an exponentially decreasing envelope [53], called inverse sweep (see Figure 4.2).

The measured impulse response $h_m(t)$ can be derived from

$$h_m(t) = \underbrace{s(t) * c(t)}_{\delta(t)} * h(t) + n(t) * c(t) \ . \tag{4.2}$$

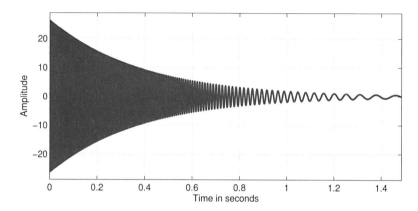

Figure 4.2.: The inverse sweep is the deconvolution signal for the exponential sweep with constant envelope. The envelope is not constant but exponentially decreasing.

The convolution of the excitation signal $s(t)$ and deconvolution signal $c(t)$ ideally results in a Dirac pulse, and thus the impulse response $h(t)$ of the system is obtained. The additive noise term is composed by the recorded background noise convolved with the inverse sweep.

The upper left part of Figure 4.3 shows a spectrogram (frequency on the vertical axis and time on the horizontal axis) of the impulse response without impulsive noise. The impulse on the left (red part) decays into the Gaussian background noise (blue part).

The upper right side of Figure 4.3 shows the contribution of the impulsive noise, which is expressed by the second summand in Eq. (4.2). Impulsive components in the recorded noise signal $n(t)$ will result in delayed representations of the deconvolution signal (the inverse sweep) in the measured impulse response.

A measurement disturbed by impulsive noise is now simulated by superpositioning both components (as shown in the upper part of Figure 4.3). This simulation is demonstrated in the lower part of Figure 4.3.

The deterministic structure of the interference in the impulse response is more

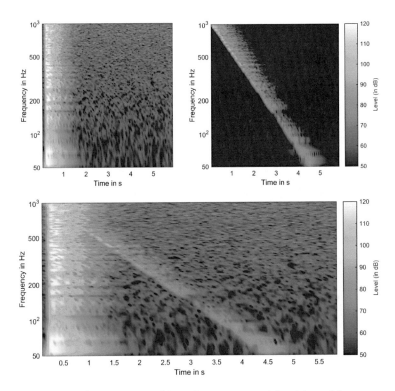

Figure 4.3.: Top: Spectrogram of impulse response without impulsive component on the left and inverse sweep caused by impulsive noise on the right side. Bottom: Impulse response including the inverse sweep caused by the impulsive noise.

problematic than the typical statistic random (stationary and broadband) noise contained in the blue part of the spectrogram. Figure 4.5 shows the reverberation time T_{20} evaluated for the undisturbed (blue) and the disturbed impulse response (green). For the room acoustic parameter, a noise compensation is applied, which includes a truncation of the RIR at the point where the signal decays into the constant noise floor (c.f. Chapter 3). This technique is able to remove the impact of the impulsive noise for low frequencies (in this example for < 400 Hz), due to sufficient time and level distance from the decaying impulse response. This can be further interpreted by looking along the horizontal (constant frequency) impulse response parts (red) decaying into the noise part (blue) without signal and noise overlap. The usage of a robust noise compensation technique for the room acoustic parameter calculation that includes a truncation of the impulse response limits the evaluation errors on the directly affected frequencies, where signal and disruption overlap. For no or other noise compensation techniques, the lower frequencies will also be affected.

Figure 4.4 illustrates two third octave bands of the energy room impulse response. For the 100 Hz band, the part of the inverse sweep occurs clearly separated from the signal part (Figure 4.4 top) and is eliminated automatically by the noise compensation method.

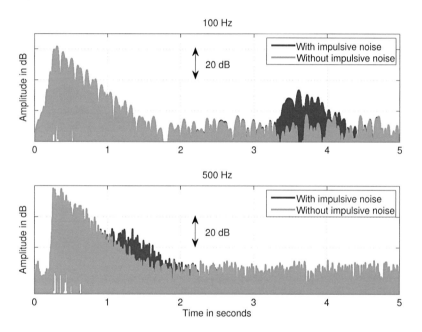

Figure 4.4.: The 100 Hz third octave band (top) and the 500 Hz third octave band (bottom) of the impulse response. The parts of the inverse sweeps appear at different positions in the impulse response and therefore have different impacts on the evaluated reverberation time and other room acoustic parameters.

For higher frequencies, however, both components overlap and cannot be separated. Around 500 Hz, the inverse sweep seems to extend the decay of the room (Figure 4.4 bottom), which results in an overestimated reverberation time (Figure 4.5). The error characteristic has various influence factors: The occurrence of the impulse during the measurement determines the occurrence of the inverse sweep in the RIR and therefore the frequency at which both components overlap. The amplitude of the inverse sweep and the related size of the error are proportional to the amplitude of the impulsive noise. Since the envelope of the inverse sweep is not constant (see Figure 4.2), an overlap at high frequencies results in larger errors.

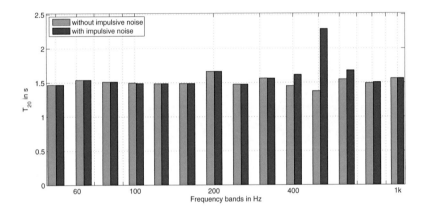

Figure 4.5.: Evaluated reverberation time T_{20} from the simulated impulse response without (green) and with (blue) impulsive noise. This comparison helps to show the effects and is not available in real measurements.

4.2. Detection of Impulsive Noise

The automatic and reliable detection of the occurrence of an impulsive noise event during a measurement is not trivial. The visibility of the distortion in the final impulse response depends on the level of the impulsive noise, compared to the level of the excitation signal, and on the time when the noise occurs.

In Figure 4.6, the broadband impulse response of the simulated example is shown, which is evaluated by using Eq. (4.2). The distortion is visible only in direct comparison to the version without impulsive noise. An automatic detection is not possible in this representation, even though the impact on room acoustic parameters is obvious (c.f. Figure 4.5) for the simulation approach where the noiseless version is known. For real measurements, the effect would be similar, but it is not possible to identify it as an error.

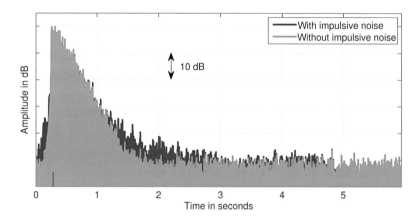

Figure 4.6.: Energy room impulse response with and without an impulsive
noise event during the measurement.

A direct detection of the impulsive noise using the sound pressure level of the recorded microphone signal (before deconvolution) is also very difficult. Despite the constant envelope of the excitation signal, the frequency-dependent transfer functions, which are mainly caused by the room (and the loudspeaker), lead to strong fluctuations in the envelope of the recorded signal. These temporal fluctuations make it almost impossible to detect impulsive noise reliably. Figure 4.7 (top) shows the small differences between the time signals caused by the impulsive noise. The impulsive noise is also plotted in yellow for illustration purposes. In the spectrogram of the recorded signal (Figure 4.7 bottom), the impulsive noise can be visually detected more easily. In this representation, image processing algorithms applied to the spectrogram might be able to detect the impulsive noise as a vertical edge. However, the required signal processing is different from the acoustic one and it is difficult to acoustically justify and interpret the thresholds, parameters, and results. The detection method proposed in the following chapter, on the contrary, is based only on ordinary acoustic signal processing and provides physically easy interpretable results (i.e. the estimated background noise).

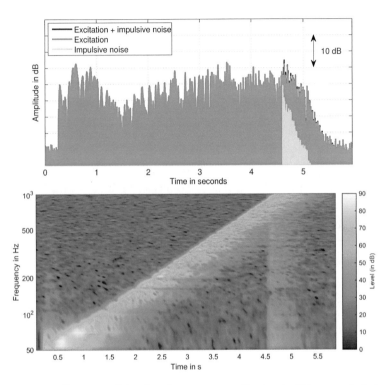

Figure 4.7.: Time signal (top) and spectrogram (bottom) of the recorded signal in the room with an impulsive noise event during the measurement at t = 4.5 s.

4.2.1. Proposed Detection Algorithm

The dominance of the excitation signal in the recorded microphone signal impedes the detection of impulsive sounds in the recorded signal. Since the excitation signal and at least an estimation of the measured system are known, it is however possible to subtract the signal part from the recorded sequence to obtain an estimation of the background noise.

The best available estimation of the true impulse response $h(t)$ is $h_e(t)$, the early part of the measured impulse response $h_m(t)$ where the signal protrudes above the noise floor. The estimated impulse response $h_e(t)$ can be obtained by applying a time window $w(t)$ to the measured impulse response:

$$h_e(t) = h_m(t) \cdot w(t) \; . \tag{4.3}$$

Here and in the following the index, "m" refers to a measured signal (including noise components) and "e" refers to estimated signals or systems. The optimal window $w(t)$ fades in at the delay of the system when the impulse response first raises above the noise level, and fades out at the intersection time where the decay of the system intersects with the background noise (see Figure 4.8). Lundeby et al. [5] developed an iterative algorithm to detect the intersection time automatically. As this algorithm has proven to give reliable results even for non-stationary noise, it is therefore used in this study.

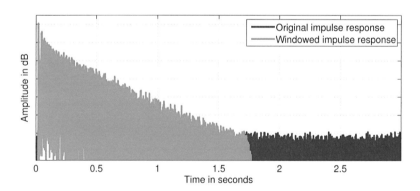

Figure 4.8.: Measured room impulse response before and after windowing (c.f. Eq. (4.3)).

The presented approach starts with the recorded signal at the microphone position, which is the response of the room to the excitation signal. The estimated excitation signal recorded at the microphone position, $r_e(t)$ is

$$r_e(t) = h_e(t) * s(t) . \tag{4.4}$$

Since the actual recorded signal $r_m(t)$ is the summation of the room response to the excitation signal and the background noise (Equation (4.1)), it is possible to estimate the background noise $n_e(t)$ by subtraction:

$$n_e(t) = r_m(t) - r_e(t) . \tag{4.5}$$

To simplify this calculation, the noise can be estimated by muting the signal part of the measured impulse response and convolving the result with the excitation signal:

$$n_e(t) = [h_m(t) \cdot (1 - w(t))] * s(t) . \tag{4.6}$$

The effect of this subtraction technique can be illustrated as follows: The group delay of the sweep increases monotonically from zero, for the start frequency f_1, to the length of the sweep, for the stop frequency f_2. A convolution with the sweep applies a frequency-dependent time shift according to this group delay. In this case, the muting time window of Equation (4.6) is shifted in time in dependence of the frequency (see Figure 4.9). This way the excitation signal including the room decay is removed from the recorded signal.

The success of this procedure depends on a reliable estimation of the system response $h_e(t)$. Estimation errors limit the performance. The true impulse response $h(t)$ continues after the truncation point. This results in remaining signal parts in the estimated background noise. The second error is noise that is included in the nominally clean response, despite windowing. This part of the noise cannot be separated from the measured impulse response and will be missing in the estimated background noise. Due to these limitations, a subtraction of the estimated noise from the recorded signal does not provide an

71

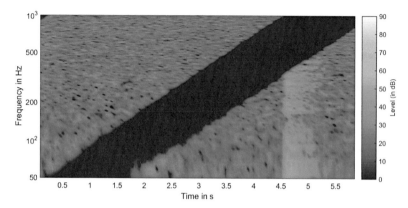

Figure 4.9.: Spectrogram of the estimated background noise. The muting time window of Eq. (4.6) is shifted in time according the frequency dependent group delay of the sweep.

advantage. (The subtraction will result in the measured impulse response with a fade-out at the intersection time.) For this reason, perfect noise compensation cannot be achieved with this technique.

Nevertheless, this background noise estimation technique is suitable to detect impulsive noise. It is possible to listen to the background noise estimation and to use a human decision when an impulsive event is present. This procedure is appropriate for single measurements, or for the identification of the noise's sources. For a larger number of measurements, an automatic assessment is needed.

A definite decision, whether impulsive noise is present or not, is difficult. The transition from stationary noise with many impulses of similar amplitude to an impulsive event that clearly stands out is smooth. The essential question is up to which level one impulsive event has to protrude over the others to have a significant effect for further usage. Figure 4.10 shows the estimated background noise for the example RIR. In the estimated noise $n_e(t)$, the impulsive sounds protrude clearly above the stationary background noise and can easily be detected. One simple measure to characterize the impulsive noise in the measurement is $L_{\max,\mathrm{rms}}$, the logarithmic ratio of the maximum amplitude of the background noise and the root mean square (also known as crest factor):

$$L_{\text{max,rms}} = 20 \cdot \log_{10} \frac{\max |n_e(t)|}{\sqrt{\frac{1}{t_{\text{IR}}} \int\limits_0^T |n_e(t)|^2 \, \mathrm{d}t}} \, . \tag{4.7}$$

For ideal Gaussian distributed stationary noise free of impulses $L_{\text{max,rms}} = L_{\text{stat}} \approx 12...14$ dB, depending on the number of measured samples. Additional impulsive components in the estimated background noise will cause a greater maximum value, while the root mean square only increases slightly. If the maximum pressure is twice the theoretical value (+6 dB), the ratio $L_{\text{max,rms}}$ exceeds 20 dB and a significant effect of the disruption on the measurement can be assumed. For calculated values of $L_{\text{max,rms}}$ around or slightly above L_{stat}, the background noise shows stationary behavior.

Figure 4.10 shows both estimated background noise signals (with and without an impulsive component) calculated with Equation (4.6). The calculated $L_{\text{max,rms}}$ values are also marked in the figure. With an impulsive component, the ratio $L_{\text{max,rms}} = 24$ dB is significantly higher than the theoretical value L_{stat} for stationary noise. Without an impulsive component, the ratio $L_{\text{max,rms}} = 13$ dB is in the range of stationary noise. Comparing the calculated $L_{\text{max,rms}}$ to the theoretical L_{stat} for stationary noise is a suitable indicator for the presence of impulsive noise components within the background noise.

In the next section, two examples of real measured rooms are shown to demonstrate the procedure and the results under realistic conditions.

4.3. Experimental Validation

The excitation signal and the impulsive noise are recorded separately using the same equipment and the same room. The noise is inserted into the measurement, subsequently allowing a detailed comparison of all processing steps, comparing the signals with and without impulsive noise components. The complete measurement

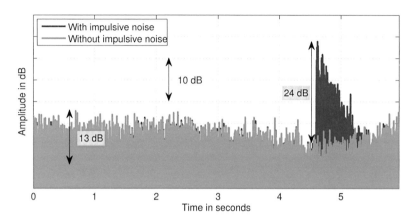

Figure 4.10.: Estimated background noise with (blue) and without (green) impulsive noise according to Eq. (4.6). The Max-RMS ratio is 13 dB without and 24 dB with the impulsive distortion.

and room acoustic evaluation is compliant with the ISO 3382 standard [3] and has been performed with the ITA-Toolbox [11].

4.3.1. General Assembly Hall

The first example is the general assembly hall of RWTH Aachen University (for detailed information see Appendix A). The dodecahedron loudspeaker was positioned at the stage, and the microphones were located in the audience area. The impulsive noise was produced by a coughing person at a different part of the audience area.

Figure 4.11 shows a comparison of the recorded signal at the microphone position with (blue) and without (green) impulsive noise. The difference is hardly visible in the envelope. To clarify the position and the amplitude of the disruption, the impulsive noise is plotted separately.

For this scenario, the influence of the impulsive disruptor can be identified clearly at the end of the room impulse response, only in comparison to the clean version

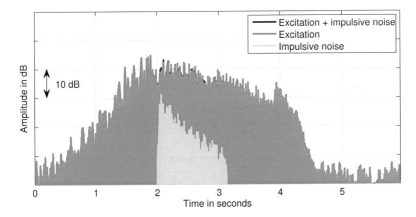

Figure 4.11.: Recorded signal at the microphone position with (blue) and without (green) impulsive noise. The difference of both signals is hardly visible. The impulsive noise is plotted separately for illustration purpose.

that includes only stationary noise (see Figure 4.12). A part of the inverse sweep clearly protrudes above the noise floor.

Figure 4.13 shows the spectrogram of the impulse response. The low frequency part of the inverse sweep (50 - 400 Hz) overlaps with the impulse response from $t = 0$ s to $t \approx 2$ s (the time of the occurrence of the impulse in the measurement). This part causes large errors in the room acoustic analysis, as seen in Figure 4.14, although this part of the disruption can hardly be detected in the spectrogram.

The second part of the inverse sweep (above 400 Hz) appears before $t = 0$ and is usually cut away by the deconvolution process.

Figure 4.15 shows the estimated background noise for the measurement with and without the impulsive noise component. Now the impulse is clearly visible and the $L_{\mathrm{max,rms}}$ value of about 26 dB indicates a significant disruption. For the case without impulsive noise, the Max-RMS ratio is only 12 dB, which is in the range for stationary noise.

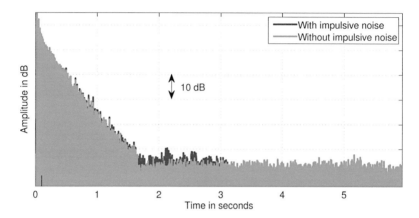

Figure 4.12.: Energy room impulse response of the assembly hall with (blue) and without (green) impulsive noise. The difference can hardly be observed in the middle where the inverse sweep slightly protrudes over the background noise.

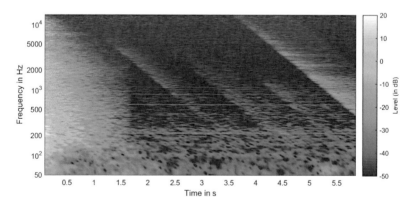

Figure 4.13.: Spectrogram of disturbed impulse response of assembly hall.

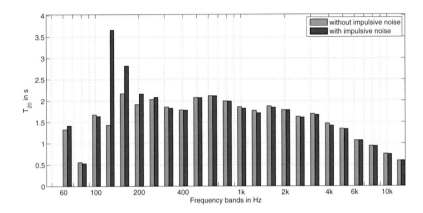

Figure 4.14.: Evaluated reverberation time for the assembly hall measurement. The error occurs only below 400 Hz where signal and noise overlap. This comparison helps to show the effects and is not available in real measurements.

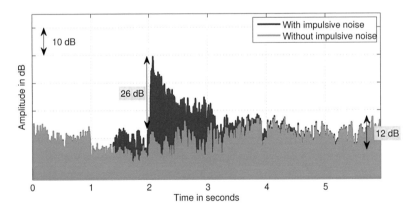

Figure 4.15.: Estimated background noise of measurement with (blue) and without (green) impulsive noise according to Eq. (4.6). The Max-RMS ratio increases from 12 dB to 26 dB with the impulsive noise component.

4.3.2. Ordinary Room

The second example is a measurement of an ordinary room. The dimensions of the room are 7 x 4 x 2.8 m^3. Source and receiver were positioned in different parts of the room. The impulsive noise was produced by dropping a pen from a height of 1.2 m in the middle between the loudspeaker and microphone position. In this example, the impulsive noise occurs at the end of the stop margin (see Figure 4.16). The stop margin is the silent part of the excitation signal after the sweep, when the recording is still continued. This part ensures that the last excited frequency components can decay into quiet before the recording period stops.

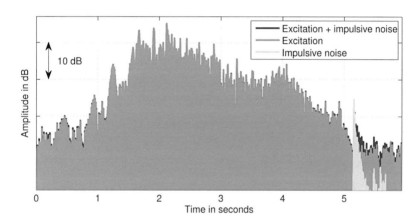

Figure 4.16.: Logarithmic level of the signal recorded at the microphone position. The impulsive noise occurs in the stop margin after the excitation signal at around 5.6 s.

Impulsive noise often occurs during the stop margin, due to noise produced by humans. They often assume that the measurement is already finished and disturb the measurement for example by footsteps. For the case that the impulsive noise occurs during the stop margin, the inverse sweep in the RIR will never overlap with the direct sound of the impulse response (see Figure 4.17). Nevertheless, the overlap of the decaying part of the room impulse response with the inverse sweep can significantly affect the room acoustic parameter evaluation. Figure 4.18

shows that the error of the reverberation time T_{30} above 2 kHz rises up to more than 100%.

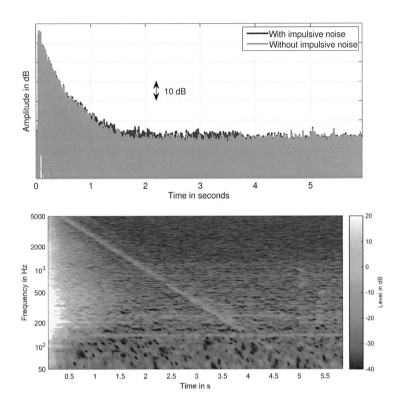

Figure 4.17.: Time signal (top) and spectrogram (bottom) of the disrupted energy room impulse response. The influence of the impulsive noise in the time signal is very small. The inverse sweep can be seen as a diagonal line in the spectrogram from 5 kHz at 0.5 s to 100 Hz at 5 s.

In the estimated background noise, shown in Figure 4.19, the impulsive noise can easily be identified. The calculated Max-RMS ratio of $L_{\mathrm{max,rms}} = 23$ dB clearly identifies an impulsive component. In comparison, the noiseless version show a value of $L_{\mathrm{max,rms}} = 13$ dB.

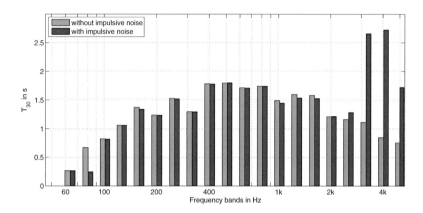

Figure 4.18.: Evaluated reverberation time for the ordinary room without (green) and with (blue) impulsive noise. This comparison helps to show the effects and is not available in real measurements.

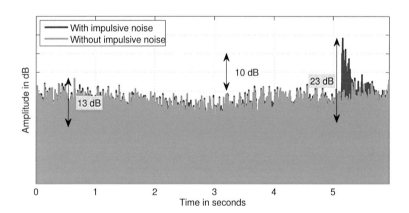

Figure 4.19.: Estimated background noise of the ordinary room measurement according to Eq. (4.6). The $L_{\mathrm{max,rms}}$ ratio increases from 13 dB without impulsive noise to 23 dB if impulsive noise is present.

4.4. Summary and Conclusion of Impulsive Noise

A method to identify impulsive noise in measurements is proposed. The motivation of this work is to overcome a major disadvantage of sweep measurements, which is the sensitivity to impulsive noise. Nevertheless, the detection algorithms can be applied to any other excitation signal.

This algorithm estimates and removes the excitation signal component from the recorded signal and supplies the estimated background noise. Based on the estimation, simple quantitative impulsive noise detection is possible. The ratio of the maximum level of the background noise and the mean level is compared to the theoretical value of stationary noise. This technique allows a robust detection of impulsive noise, even if the disruption is not detectable in the room impulse response or the recorded signal. It is important to detect impulsive noise, because it still has an influence on the room acoustic parameter evaluation. The algorithm is fully automated and does not need any manual adjustment. This allows an automatic analysis directly after the measurement and a repetition of the measurement, if necessary. Furthermore, it is possible to listen to the estimated background noise to allow an individual analysis.

5

Time Variances

Room acoustic measurements are based on the assumption that the measured system is time-invariant (c.f. Section 2.1). However, in reality this hypothesis is only correct up to a certain point. In this chapter three major effects that cause time variances in room acoustic measurements are studied closely: air movement, temperature and humidity changes, and disturbing scattering objects in the room. For the temperature and relative humidity investigations, it is distinguished between changes within one measurement and changes between measurements.

For this purpose, special measurement sessions in real auditoria have been conducted that allow a detailed analysis of the different influence factors. Therefore, this chapter focuses on room acoustic parameters as target quantities. In the discussion of each section, the practical relevance of the investigated effects is discussed.

5.1. Air Movement

In practice, many sources of air movement are possible, such as moving persons or objects in the room, air draughts due to open windows or doors, or an active ventilation system. This section only investigates the effect of the air ventilation system on the measurement uncertainty, since it is assumed that other possible sources introduce less air movement to a smaller part of the room. In addition, the usage of the ventilation system has the advantage of reproducible results and easy realization of the measurements, due to the possibility to program the system.

An air draft might also include the effects of temperature and humidity change. This influence is discussed separately in Section 5.2. The most possible air movement is produced from a ventilation system and it is therefore the worst case scenario. Other sources, such as the movement of persons, are assumed to have smaller impact. The results presented in this section are partly based on results published in Guski and Vorländer [54].

5.1.1. Measurement Setup

Measurements have been conducted in the General Assembly Hall of RWTH Aachen University (for detailed information see Appendix A). The loudspeaker has been positioned on the stage and the microphones have been placed in the audience area in three parallel rows. The source has been set to a high output level to be able to keep the length of the excitation signal short (5 seconds). This ensures that within one measurement the conditions can be considered as constant. At the same time, the output level of the loudspeaker has been kept sufficiently low to ensure that nonlinearities have no influence.

The ventilation slots are arranged around the audience area (see Figure 5.1). The microphones have been positioned in different ranges to the ventilation slots to study if the effects are a function of the distance to the ventilation system.

The air conditioning system has been programmed to switch status between on and off. It was not possible to control the temperature regulation. This means that while the ventilation was active, heating or cooling was active, too. Therefore, the analysis method in this section is designed to account for the effects of temperature changes, whose influences are analyzed separately in Section 5.2. The single phases while the status of the air conditioning has been constant are four hours long to allow the system to reach a stationary condition. The phases have been alternatingly repeated (see Figure 5.2) to facilitate the identification of a correlation between the room acoustic parameter and the state of the ventilation system. Other long-term effects (such as heating-up of the equipment or changes

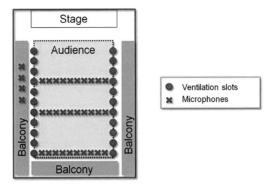

Figure 5.1.: Site plan of General Assembly Hall of RWTH Aachen University "Aula 1" with microphone and ventilation outlet positions.

of background noise level) will follow a different pattern and can therefore be separated. During the measurement period of about 24 hours, no persons have been present in the auditorium. The first hours after the last persons have left the room are excluded from the measurements, allowing the room to settle. Every 45 seconds a transfer function measurement has been started and stored automatically, leading to more than 1900 measurements in total.

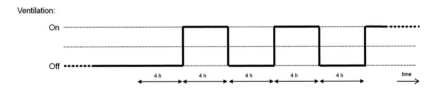

Figure 5.2.: Schedule of the air conditioning system. After a settling time of a few hours, the status of the ventilation system is switched on and off every four hours.

Eight temperature and three humidity sensors have been distributed over the audience area. The temperature and the relative humidity have been measured immediately after each acoustic measurement to allow a detailed analysis afterwards. The measurements and the equipment conform to the requirements of ISO 3382 (see Chapter 2.1.2 for more details).

5.1.2. Analysis of Measurements

A detailed analysis of the time structure of the room impulse responses does not show any differences between the phases of the ventilation system. An analysis of the room acoustic parameters, however, reveals a correlation. Figure 5.3 shows the evaluated reverberation time T_{20} as a function of time. The red highlighted areas mark the phases when the ventilation system is active. It can be seen that there is no constant offset in the evaluated parameter for the switch of the ventilation status. The slow variations of the reverberation time that occur over periods of hours are neglected in this section. They are discussed in detail in Section 5.2. Focusing on the fast fluctuations that occur over minutes (or from one measurement to the next), it can be seen that the variance is significantly higher for the phases with active ventilation system.

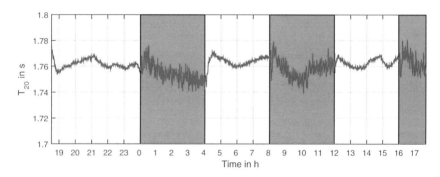

Figure 5.3.: Reverberation time T_{20} as a function of time. The phases with activated ventilation system are highlighted in red. The figure shows the 2 kHz octave band for microphone position 1.

To enable a clearer view on the short time fluctuations, a moving standard deviation window is applied to the data. This window analyzes the standard deviation using eight consecutive measurements that correspond to a measurement time of six minutes. This short window ensures that long-term changes are ignored in this analysis. One long-term effect is a temperature drift over time, since it is not possible to keep the temperature constant in a big auditorium for 24 hours while the ventilation system is switched on and off.

The result of the moving standard deviation analysis is shown in Figure 5.4. It can be seen that the relative standard deviation of the phases with active ventilation system is higher by a factor of approximately four.

The higher values of the standard deviation at the transitions where the status of the ventilation is switched are caused by the fact that the moving window contains measurements from both states. Therefore, these values should be excluded from the interpretation.

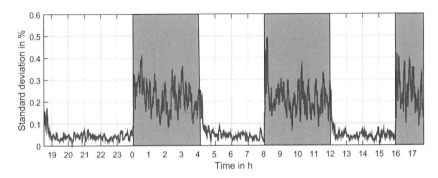

Figure 5.4.: Relative standard deviation of a moving window of reverberation time T_{20}. Window width: 6 minutes (8 measurements). The figure shows the 2 kHz octave band for microphone position 1.

Ventilation System as Noise Source

The most obvious source of the observed effect is an additional noise component for the active ventilation system. The increased background noise level reduces the peak signal-to-noise ratio (PSNR) and the random fluctuation of the evaluated parameter therefore increases if the PSNR decreases (c.f. Chapter 3). To disprove this theory, Figure 5.5 shows the measured PSNR for the exemplatively discussed microphone position and frequency band. The PSNR exceeds values of 80 dB and is (with few exceptions) constant over the complete measurement time. It can be stated that it is definitely not influenced by the status of the air conditioning system. The very high values ensure that a small change of the PSNR will not have any influence on the evaluated parameters. Hence, noise does not have to be considered as a possible source of increased variances. It is therefore most probable that the air movement in the auditorium causes the acoustic system to become instationary.

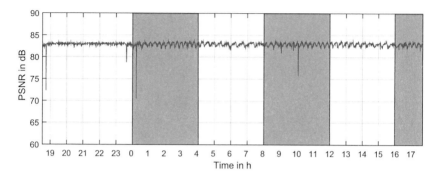

Figure 5.5.: Peak signal-to-noise ratio in dB as a function of time. No influence of the ventilation system on the PSNR is present. The figure shows the 2 kHz octave band for microphone position 1.

Evaluation Range

The dynamic range used for the evaluation of the reverberation time influences the behavior of the variances only slightly. Figure 5.6 shows a section of the measurement session for different reverberation times representing different evaluation ranges from 10 dB to 40 dB. A tendency of decreasing variances for increasing evaluation range can be observed. Larger evaluation ranges involve more averaging and small fluctuations have therefore a smaller impact. Despite the averaging, a significant difference between the phases remains visible.

Position of Microphones

The observed effects occur for every microphone position and the magnitude of the standard deviation is independent of the position. Therefore, this effect is not a function of the distance between microphone and ventilation slots. This result meets the expectations, since every reflection in the impulse response is

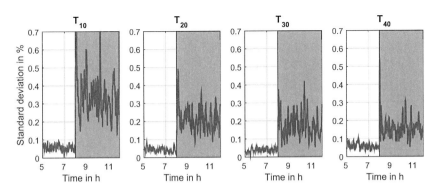

Figure 5.6.: Comparison of relative standard deviations of moving window analysis for different evaluation ranges for the reverberation times of the 2 kHz octave band.

influenced by the entire course that the corresponding sound particle travels across the room. Thus, local effects should not be visible in the room acoustic parameter.

Frequency Band

Figure 5.7 shows the moving window standard deviation analysis for four octave bands from 500 Hz to 4 kHz. It can be seen that the fluctuations increase with increasing frequencies.

Moving air in the auditorium causes small local changes of acoustic propagation characteristics. Variations in temperature, humidity, and speed of the medium cause deviations from the speed of sound c and thus changes in the time of arrival occur. Since these effects are local and different for every path of a reflection, this will cause amplification or attenuation of parts of the impulse response, depending if the interaction of different reflections are constructive or destructive. The time differences have to be referred to the period duration of the frequency to obtain the phase errors, resulting in larger errors for higher frequencies.

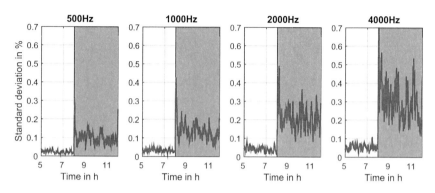

Figure 5.7.: Comparison of relative standard deviations of moving window analysis of reverberation time T_{20} for different octave band frequencies.

Energy Parameters

The energy parameters show a similar behavior. No systematic offset between an activated and a deactivated ventilation system can be observed. However, the random variances show a clear dependence on the air movement. Figure 5.8 shows the moving standard deviation analysis of the clarity index C_{80} for different octave band frequencies. The standard deviation increases by a factor of two to five when the ventilation system is active. Similar to the reverberation time, higher frequencies are more sensitive to air movement. For a smaller separation time $t_e = 50\,\text{ms}$ (used for the parameters D_{50} and C_{50}), the standard deviations are slightly higher.

5.1.3. Statistical Summary

In the previous subsection, the qualitative influence of air movement on the parameters has been shown. In this subsection, the measurement results are summarized to give a more general quantitative statement on the found empirical effects. As mentioned before, the position of the microphone has no influence on the observed effects and all channels show similar results regarding the relative standard deviation. Therefore, all measured channels are considered together.

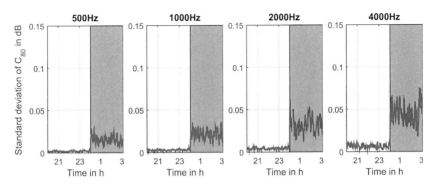

Figure 5.8.: Comparison of standard deviations of the clarity index C_{80} in dB for different octave band frequencies.

However, the frequency band and the type of room acoustic parameter are analyzed separately, since they have shown strong influence on the size of the error.

According to the *Guide to the expression of uncertainty in measurement* (GUM), the variances of N different uncorrelated uncertainty sources u_i^2 can be combined by [55, Eq. (11a)]:

$$u_c^2 = \sum_{i=1}^{N} u_i^2 \tag{5.1}$$

The single variances u_i^2 can be estimated from the squared standard deviations s^2 from a series of repeated observations (Type A evaluations). To separate the contribution of the air movement from other sources of random variations, it is distinguished between the two components air movements s_{air} and the remaining influences s_{r}. Both components can be considered as uncorrelated, since it has been shown that the additional background noise has no influence and all other conditions are kept constant. Further assuming that the sensitivity of both components is similar and equal to one ($c_i = 1$), Eq. (5.1) can be simplified to

$$s_c^2 = s_{\text{air}}^2 + s_r^2 \qquad (5.2)$$

For phases with an active ventilation system, the observable random variances are a combination of both components (as described in Eq. (5.2)). For phases with disabled ventilations system, the air movement term vanishes and the variances of the observations are equal to the remaining variances s_r^2. To isolate the contribution of the air movement, the variance of the measurement with deactivated ventilation s_{off}^2 has to be subtracted from the measurement with active airing system s_{on}^2:

$$s_{\text{air}}^2 = s_{\text{on}}^2 - s_{\text{off}}^2 \; . \qquad (5.3)$$

To apply this formula to the measurement data, each parameter, frequency band, and microphone position is handled separately. In these measurement subsets, all measurements with present scattering objects are combined into one group and the measurements with absent scatter in another. For these groups the empirical standard deviations are determined and the contribution of the air movement is calculated.

Figure 5.9 shows the results of the evaluation for the reverberation time T_{20} as a function of the frequency band for all microphone positions. It can be seen, that the results are very similar for all microphone positions, with exception of the lowest frequency band. For low frequencies the variance of the air movement decreases while the remaining variances increase, mainly caused by worse signal-to-noise ratios. The dominance of the remaining variances makes a prediction of the relatively small air movement variance impossible.

Table 5.1 summarizes the empirical results for several room acoustic parameters. Therefore, the maximum deviation of all microphone positions is used as a worst-case estimation.

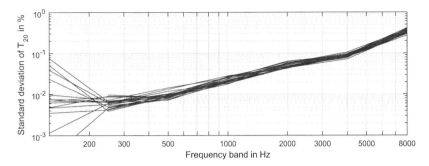

Figure 5.9.: Determined standard deviation of air movement of the reverberation time T_{20} in dependence of frequency. The different measurement positions show similar results.

5.1.4. Discussion

The investigation of the influence of the ventilation system on the room acoustic
parameters reveals that the random error is significantly larger when the ventilation system is active. This effect is more pronounced for higher frequencies and
smaller evaluation ranges of the reverberation time. The moving window standard deviation analysis shows that the standard deviation of the reverberation
time is $\leq 1\%$ in the usual frequency range. Regarded under perceptual criteria,
theses variances are small compared to the JND of 5%. The same applies to the
energy parameters. The standard deviation of the clarity index C_{80} of $\leq 0.1\,\mathrm{dB}$
is small compared to the JND of $1\,\mathrm{dB}$. For measurement applications where the
detailed phase information of several impulse responses or transfer functions
is important (i.e. superpositioning approaches), the growth in the temporal
fluctuations might have a larger influence. The concrete influence on the results
is dependent on the specific procedures and has to be evaluated separately.

| Frequency | EDT | T10 | T20 | T30 | T40 | D50 | C80 |
Hz	%	%	%	%	%	%	dB
63	1.864	2.786	0.684	1.408	3.433	0.150	0.018
125	0.286	0.149	0.275	1.228	0.613	0.113	0.007
250	0.190	0.184	0.098	0.084	0.063	0.248	0.011
500	0.182	0.188	0.120	0.074	0.073	0.373	0.016
1000	0.312	0.319	0.168	0.119	0.127	0.495	0.025
2000	0.379	0.424	0.253	0.187	0.150	0.854	0.046
4000	0.501	0.588	0.326	0.230	0.204	1.239	0.071
8000	1.020	1.156	0.640	0.449	0.359	1.521	0.119

Table 5.1.: Evaluated standard deviations caused by the air ventilation system in dependence of the frequency.

5.2. Inter-Measurement Temperature Changes

Room acoustic measurements are often made between two regular uses of the auditorium. In this time the air conditioning system is usually not activated. Hence, the temperatures during the measurements can vary strongly, depending on the weather conditions outside. Temperature ranges from $15\,°\text{C}$ to $30\,°\text{C}$ can be reached easily. The meteorological conditions of air have a large influence on the acoustic conditions, such as the sound velocity and the air attenuation (c.f. Section 2.2). These parameters, in turn, cause a change in the measured impulse responses and the evaluated room acoustic parameters.

In this section so-called *inter-measurement* changes of the acoustic conditions are discussed that occur only between and not during a measurement. Therefore, each single measurement for its own is correct and represents the system for the current acoustic conditions. It is investigated how large the influence of temperature and humidity changes on the room acoustic parameters is and if this effect can be predicted and corrected by the given theoretical relations. The question to be answered is, for example, if measurements made for $17\,°\text{C}$ and $25\,°\text{C}$ can be compared directly or if the temperature effect has to be compensated. A further question is how reliable this compensation is.

Section 2.2 describes the influence of temperature Θ and relative humidity φ on the speed of sound c and the air attenuation m. Inserted into the Eyring equation (Eq. (2.17)) the change of the reverberation time can be predicted:

$$T = \frac{24 \cdot \log_e(10)}{c(\Theta, \varphi)} \frac{V}{A - 4V \cdot m(\Theta, \varphi)} \; . \tag{5.4}$$

where the equivalent absorption area is $A = -S \cdot \log_e(1 - \bar{\alpha})$. The change in temperature can be interpreted as the stretching or compression of the time axis of the room impulse response.

In this chapter long-term measurements have been continuously performed on the temperature and the relative humidity changes. After each measurement, the temperature and the relative humidity in the room have been recorded. The eight temperature sensors were spread over the audience area analogous to the microphones.

The equivalent absorption area A is determined for every frequency band by an average of all measurements in the room. The calculation of the equivalent absorption area also uses the relation of Eq. (5.4) and thus it seems questionable to use the same relation in the calculation that should be investigated. However, since the same A is used for all measurements, an error in A is mainly just able to introduce an offset to the predicted results. The same applies to an error in the correct determination the volume of the room.

The results presented in this section are partly based on results published in Guski and Vorländer [54].

5.2.1. Eurogress Measurements

The *Eurogress Aachen* is a multipurpose hall and is used for concerts, congresses, and conferences. It has a volume of $14300\,\mathrm{m}^3$ and a capacity of up to 1700 people, depending on the seating plan. The measurement session was approximately 13 hours long. During this period the temperature changed by up to $2\,°\mathrm{C}$ and the relative humidity by about $4\,\%$. According to Eq. (5.4), these variations lead to a variation of $1\,\%$ in reverberation time for the $1\,\mathrm{kHz}$ frequency band. With this prediction, and based on the assumption of an ideal sound field with infinite reflections (c.f. Eq. (2.31)), the change of clarity index C_{80} can be estimated to be $0.05\,\mathrm{dB}$. For high frequencies of the $4\,\mathrm{kHz}$ band these changes increase to $5\,\%$ in reverberation time and $0.3\,\mathrm{dB}$ in clarity. However, this can only be considered as a rough estimation, since Eq. (2.31) is based on assumptions that are often not fulfilled in reality.

If a strong reflection is located near the time limit of $80\,\mathrm{ms}$ and a temperature change causes a time stretch of the impulse response, this might result in reflections appearing on the other side of the limit, resulting in variations of C_{80} that are significantly larger than predicted by the ideal sound field theory. The sensitivity of the clarity index to temperature changes depends significantly on the positions of the reflections in the impulse response and is therefore dependent on the room and the exact positions of sender and receiver. A general prediction of the resulting changes in C_{80} is not possible.

The temperature and humidity profile can be seen in Figure 5.10. The air conditioning was deactivated at the beginning and was switched on automatically in the later part of the measurement session, at around 7 o'clock. The first hour after the last person has left, the room has been excluded from the measurements to allow for the room to settle from the influence introduced by the measurement crew.

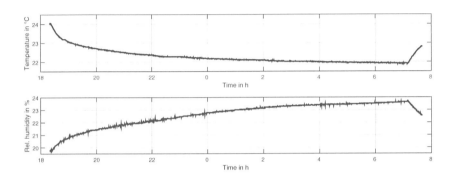

Figure 5.10.: Temperature (top) and relative humidity (bottom) profile for the measurement session in Eurogress hall.

Reverberation Time

In Figure 5.11 three examples of measured and predicted reverberation time T_{30} are shown. All examples represent reverberation times for the octave band of 1 kHz. These microphone positions have been selected to show the existing diversity of the results. For channel 6 (Figure 5.11 top), the prediction and the measurement agree with each other. The calculated correlation coefficient of both curves is $\rho = 0.92$. For channel 8 (Figure 5.11 center), the reverberation time variation differs strongly. The theory predicts a small increase of the reverberation time, whereas the measured T_{30} values initially increase strongly for the first 3 hours and then decrease for 19 hours. This uncorrelated behavior is reflected in the correlation coefficient of $\rho = -0.19$. The third example is microphone position 12 (Figure 5.11 bottom). The measured reverberation time decreases while the theoretical increases. The complete inverse behavior leads to a correlation coefficient of $\rho = -0.94$.

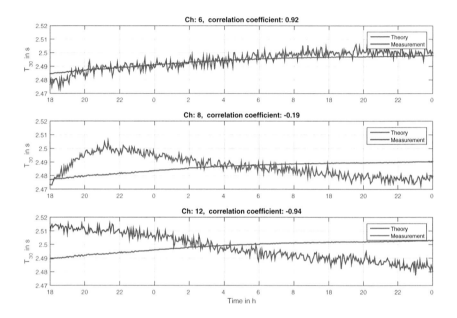

Figure 5.11.: Comparison of measured and predicted reverberation time T_{30} in the 1 kHz octave band. Three microphone positions show the diversity of the results: good agreement (top), no correlation (center), and inverse relation (bottom).

To analyze these different behaviors in detail, the correlation coefficients for all microphone positions and frequency bands are shown in Figure 5.12. For high frequencies including the 2 kHz octave band, the correlation between measurement and prediction is very high (the mean value is ≥ 0.95). This effect can be traced back to the high influence of the air attenuation constant for these frequencies. The mid frequencies from 250 Hz to 1 kHz show very diverse results, covering everything between very good correlations such as $\rho = 0.96$ and negative correlations up to $\rho = -0.99$. For the lower frequency bands of 63 Hz and 125 Hz, the correlation coefficients approach zero. The lower signal-to-noise ratio for lower frequencies introduces a random fluctuation of the evaluated reverberation time that is dominant towards the small influence of the temperature change. No single channel can be identified that shows neither very good nor very poor correlations throughout the medium frequency range.

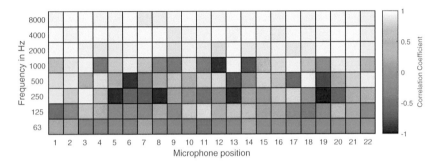

Figure 5.12.: Correlation coefficient between measured and predicted reverberation time T_{30} depending on microphone position and frequency band for Eurogress Aachen measurements.

Clarity Index

A comparison between theory and measurement of the clarity index C_{80} is shown by three different examples in Figure 5.13. The theoretical changes of C_{80} are determined by theoretical changes in the reverberation time T and the assumption of a diffuse sound field (c.f. Eq. (2.31)). This rough estimation gives the same behavior for the complete room, whereas the real clarity index is dependent on the position in the room. This is why there is a constant offset for every position and frequency band. To simplify the comparison of theory and measurement, the mean values of both curves are subtracted in the plots. Figure 5.13 (top) shows relatively large changes in C_{80} for the 4 kHz octave band. Theory and measurement agree well, however only a small number of frequency-microphone combinations show similar behavior. For the second example (Figure 5.13 center) of the 1 kHz octave band, the changes in the measurements are clearly larger but the tendency is equal. While the theory predicts a drop of around 0.03 dB, the measurements vary in a range of more than 0.2 dB. In the third example (Figure 5.13 bottom), measurement and theory show completely different behavior. Similar to the results of the reverberation time, there is no correlation between the goodness of the prediction and the position of the microphone or the frequency band.

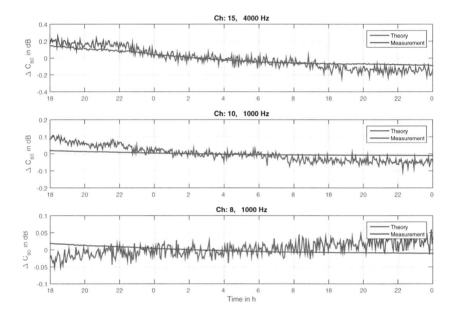

Figure 5.13.: Comparison of measured and predicted clarity index C_{80} in the 4 kHz and 1 kHz octave bands. Three microphone positions show the diversity of the results: good agreement (top), underestimated effect by theory (center), and inverse relation (bottom).

5.2.2. Assembly Hall Measurements

A second measurement session in another auditorium and with a different temperature profile has been conducted. The measurements have been performed in the General Assembly Hall of the RWTH and the status of the air conditioning system has been switched between on and off every four hours (c.f. Section 5.1).

Figure 5.14 shows the relative humidity and the temperature profile of the measurements with red highlighted areas for active air conditioning. The temperature decreased with active conditioning system because the measurements have been conducted during a warm summer day and the system has been in cooling mode. The temperature has changed within a range of 3.2 °C and the relative humidity within a range of 7 %. According to Eq. (5.4), these variations lead to variation of the reverberation time of 1.3 % at 1 kHz. The theoretical change of the clarity according to Eq. (2.31) is 0.08 dB. Again, the real occurring changes are dependent on reflections around 80 ms and are expected to differ strongly from the prediction.

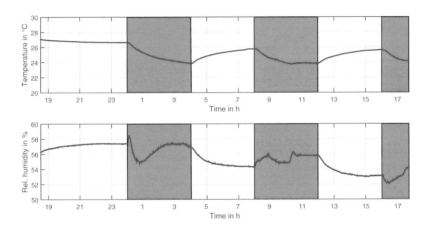

Figure 5.14.: Temperature (top) and relative humidity (bottom) during the measurement session in the General Assembly Hall. The red highlighted areas indicate the activity of the air conditioning system.

Reverberation Time

The results confirm the findings from the Eurogress Aachen measurements. Figure 5.15 shows three demonstrative comparisons between measurement and theory to show the diversity. The results belong to the same 1 kHz octave band and represent only different microphone positions. The examples cover a range of good agreement (Figure 5.15 top), over no correlation (Figure 5.15 center) up to inverse behavior (Figure 5.15 bottom).

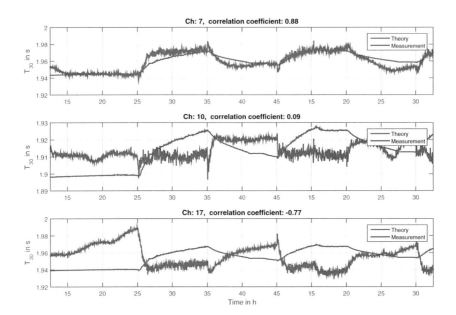

Figure 5.15.: Comparison of measured and predicted reverberation time T_{30} in the 1 kHz octave band. Three microphone positions show the diversity of the results: good agreement (top), no correlation (center), and inverse relation (bottom).

An overview of the frequency and microphone position influence on all correlation coefficients is given in Figure 5.16. Again, there are no microphone positions that stand out with overly good or bad correlation. A tendency for the dependence on the frequency can be recognized, where both results correspond better for higher frequencies. However, this effect is less pronounced than for the Eurogress measurements (c.f. Figure 5.12). This might be due to the smaller volume of the assembly hall ($5500 \, \text{m}^3$) compared to the Eurogress ($14300 \, \text{m}^3$) and the therewith related smaller influence of the air attenuation.

Furthermore, the temperature profile of Eurogress measurements is rather trivial. The temperature mainly increases over the complete measurement period. Other long-term effects that are constant over a long time easily result in an incorrect high positive or negative correlation. The rather complex temperature profile of the General Assembly Hall makes that kind of false identification of other influences very unlikely.

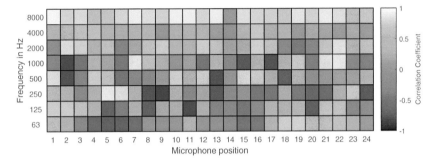

Figure 5.16.: Correlation coefficient between measured and predicted reverberation time T_{30} as function of microphone position and frequency band for the measurements in General Assembly Hall.

Clarity Index

The variations of C_{80} for the General Assembly Hall are clearly larger than the predicted values. Figure 5.17 (top) shows the typical example: The tendencies of measurement and simulation agree, but the magnitude of the measurement results is significantly larger. Whereas, the prediction varies in a range of 0.05 dB, the measurement values are spread over a range of 0.45 dB. Figure 5.17 (center) shows an inverse behavior. When the theory predicts a decreasing clarity index, the measured C_{80} values increase. The magnitude of the change is again significantly higher in the measurements (range of 0.09 dB for the theory versus 0.48 dB for the measurements). Figure 5.17 bottom shows one of the few examples where the magnitudes of variation fit quite well.

103

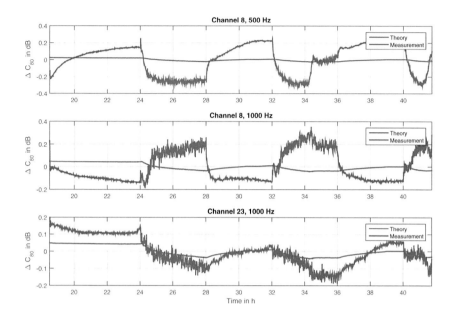

Figure 5.17.: Comparison of measured and predicted clarity index C_{80} in the 500 Hz and 1 kHz octave bands. Three microphone positions show the diversity of the results: same tendencies but higher magnitudes of the measured values (top), inverse behavior (center), and relative good agreement (bottom).

5.2.3. Detailed Analysis of Clarity Index

To investigate the sources of the large and heterogeneous results the of clarity index, three impulse responses of the measurement session are selected and analyzed in detail. Figure 5.18 shows a section of the impulse responses around the 80 ms time limit for the same octave band and microphone position for three different temperatures: 26.6 °C, 26.4 °C and 25.5 °C.

First, it can be seen that there is a reflection at the 80 ms limit. The appearance of the reflection changes from before the 80 ms (for the first measurement) to after the limit (for the last measurement). However, these circumstances always

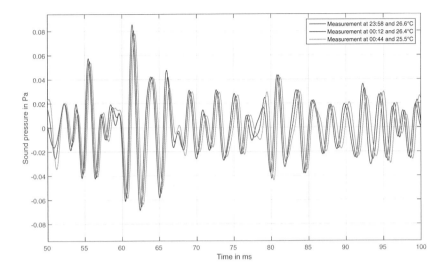

Figure 5.18.: Time structure of an impulse response measured for three different temperatures. The position of one reflection changes due to the time stretching from before the time limit of 80 ms to after it. Furthermore, interference effects cause increasing (i.e. at 51 ms) and decreasing amplitudes (i.e. at 92 ms) can be observed.

result in the same relation between temperature and clarity index: Decreasing temperatures lead to deceasing C_{80}. The impact results in the same kind of error as the time stretching theory (c.f. Eq. (2.31)) and can therefore only enhance this effect but cannot explain the occurring inverse character between prediction and measurement.

Furthermore, the amplitude of the reflection seems rather small. To analyze the possible effect of a movement of one reflection, the second impulse response is investigated more closely. The temperature range of 3.2 °C in the measurements causes the speed of sound to change by 2.1 m/s. And this, in turn, leads to a maximum time stretch or compression of around ±6 ms. Instead of scaling the time axis, the effect of the temperature change can also be analyzed by shifting the 80 ms time limit by ±6 ms. Figure 5.19 shows the difference in the calculated clarity index regarding the value for 80 ms. Since the change is not symmetrical for positive and negative temperature changes, the plot shows both scenarios. For a temperature decrease of $3.2\,^\circ$C, the clarity index C_{80} would decrease by about -0.25 dB. This effect alone is not able to explain the complete variations in the measurements of 0.45 dB, but it shows that this effect is significantly larger than the predicted change of the diffuse field theory (0.05 dB). For a positive deviation of $3.2\,^\circ$C, the difference in C_{80} would be even larger and reach 0.66 dB.

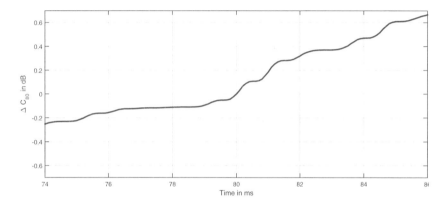

Figure 5.19.: Sensitivity of clarity index C_{80} of a real impulse response to time limit changes. The change of the limit of ± 6 ms has the same effect as time stretching of the impulse response temperature caused by a change of $\pm\,3.2\,^\circ$C.

In addition to the time differences between the three impulse responses of Figure 5.18, amplitude changes can also be observed. Some peaks of the impulse response increase over the time (for example around 51 ms and 79 ms), while others decrease (61 ms or 92 ms). The sources of the observations are changes in the phases of different propagation paths across the room. Dependent on the phase relation between two or more reflections, the resulting amplitudes can be amplified by constructive interference or be attenuated in case of destructive interference. These changes are not systematic for increasing or decreasing temperatures, but are strongly influenced by the geometry of the room, position of sender and receiver, and temperature distribution in the room. This effect is not covered by the time stretching theory and explains the significant differences in the measured values for different positions.

5.2.4. Summary

The evaluated measurement results for both sessions are summarized and compared to the theoretical predictions. Therefore, both measurement sessions are used. For the T_{20} measurements, Figure 5.20 and Figure 5.21 show an overview to illustrate the differences. Each subplot illustrates the relative change in T_{20} as a function of temperature difference. The different colors represent the microphone channels. Each row demonstrates one octave band of 250 Hz or 1 kHz. The left column shows the theoretically calculated results and the right column the evaluated measurements.

The relative humidity has also changed during the measurements. The effect on the parameters is included in the evaluated parameters and is considered in the theoretical calculations. However, all humidity values are combined for illustration purpose. An exact unique functional relation between temperature change and parameter change is only possible for one constant relative humidity value. Therefore, a part of the diversification of the measurements can be explained by the non-observance of relative humidity. However, the good congruence of the theoretical predictions shows that this part should be negligible.

In contrast to the theoretical data, the different channels of the measurement data show partly complete diverse behavior. For decreasing temperatures, for some channels positive and for others negative behavior can be observed. The number of channels showing positive and negative behavior is approximately similar for lower frequencies, but for increasing frequencies a tendency to positive deviations is obvious.

Comparing the magnitudes between theory and measurement, the previous observations for single channels are confirmed. Several channels show significantly larger deviations than predicted. The theoretical differences scale with the regarded frequency band, whereas for the measurement no clear tendency can be observed. The theoretical effect can be observed for some channels, but another larger impact often is dominant. To allow an estimation of the error sizes of both results (theory and measurement), the 95th percentile of the absolute (relative) deviation is determined from the given data. The values of the measured change in the reverberation time T_{20} are listed in Table 5.2 and the theoretical results in Table 5.3. The results for further parameters are given in Appendix C.

$\Delta\Theta$ °C	125 Hz %	250 Hz %	500 Hz %	1000 Hz %	2000 Hz %	4000 Hz %
-0.1	2.032	1.533	1.300	1.597	1.252	2.364
-0.3	1.793	1.434	1.395	1.446	1.536	2.239
-0.5	1.454	1.887	2.133	2.398	2.119	3.828
-0.7	2.038	2.508	2.682	2.322	2.510	3.283
-0.9	2.879	3.078	2.844	2.242	2.564	3.484
-1.1	3.104	3.144	2.646	2.238	3.031	3.314
-1.3	2.885	2.612	3.212	3.181	3.842	4.644
-1.5	3.251	3.307	2.894	3.644	4.965	6.119
-1.7	3.390	3.407	3.209	4.394	5.943	7.505
-1.9	3.703	3.601	3.371	5.456	7.655	9.795
-2.1	4.058	3.889	3.483	5.794	9.269	11.468

Table 5.2.: Calculated variance in measurement of T_{20} based on the meteorological conditions of both measurements. Deviations are listed depending on the temperature difference.

5.2.5. Conclusion

The theoretically calculated changes in the room acoustic parameters can also be observed in the measurements. The presented formulas can be used to predict this effect. Therefore, it is also possible to compensate these effects to allow a comparison of two measurements that were made under different conditions. For temperature differences smaller than $3\,°C$, the change in reverberation time is below 2% and can be neglected for conventional room acoustic measurements.

Nevertheless, the measurements also showed a further effect that is clearly larger in magnitude. The error size also increases with temperature difference, but the sign of the change varies randomly with the observed frequency band and the microphone position. It is assumed that this behavior results from several constructive and destructive overlappings of reflections in the RIR. These

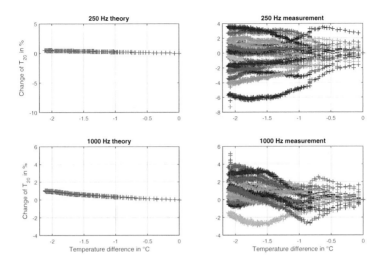

Figure 5.20.: Scattering plots showing the relation between temperature difference and change in reverberation time for the theoretical model (left) and Eurogress measurements (right). The colors represent the different channels.

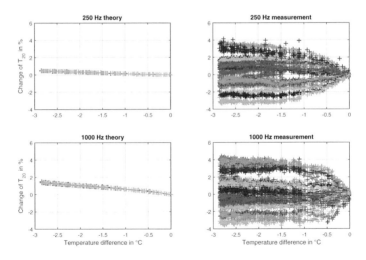

Figure 5.21.: Scattering plots showing the relation between temperature difference and change in reverberation time for the theoretical model (left) and Assembly Hall measurements (right).

$\Delta\Theta$ °C	125 Hz %	250 Hz %	500 Hz %	1000 Hz %	2000 Hz %	4000 Hz %
-0.1	0.016	0.020	0.037	0.089	0.201	0.240
-0.3	0.064	0.074	0.118	0.119	0.165	0.990
-0.5	0.112	0.128	0.289	0.280	0.345	2.159
-0.7	0.162	0.185	0.416	0.396	0.499	3.081
-0.9	0.208	0.241	0.513	0.492	0.899	3.349
-1.1	0.255	0.298	0.616	0.642	1.268	3.797
-1.3	0.300	0.353	0.709	0.884	1.814	4.137
-1.5	0.347	0.410	0.798	1.182	2.509	4.280
-1.7	0.390	0.463	0.872	1.414	3.065	4.348
-1.9	0.435	0.525	0.958	2.021	4.577	5.645
-2.1	0.478	0.579	1.024	2.429	5.587	6.950

Table 5.3.: Calculated theoretical change of T_{20} based on the meteorological conditions of both measurements. Deviations are listed depending on the temperature difference.

influences depend on the detailed geometry of the setup and the temperature distribution in the room, and are therefore too complex to predict.

From the empirical data, it can be seen that even for the highest frequencies the error of the reverberation time usually stays below 4% for temperature differences of $< 2\,°\text{C}$. This means that this effect is smaller than the perceivable difference of about 5% and is not relevant for the description of subjective parameters. For technical the error of the reverberation time propagates to the evaluated parameter and the tolerable error depends strongly on the individual purpose.

5.3. Intra-Measurement Temperature Changes

In this section, intra-measurement time variances are analyzed. Whereas, in the previous section, where each measurement by itself is a correct representation of the immediate conditions, changes of the acoustic system conditions during a measurement itself will be the basis of incorrect measurements. As already discussed in Section 2.1.3, the sweep measurements preferred nowadays are more robust concerning time variances compared to MLS, since every frequency is only excited for a very short time. Nevertheless, there are various scenarios where intra-measurement time variances become a problem: For large halls or rooms with high ambient noise levels, often very long excitation signals or averaging is used to guarantee a sufficient signal-to-noise ratio. For cathedrals, for example, excitation signal lengths of $> 90\,\text{s}$ are not uncommon. Another example are special measurement sessions where several impulse responses are set into relation to the correct phase to provide the resulting "one" measurement. The measurement time can easily last several hours. For the superposition approach to synthesize arbitrary source characteristics of spherical harmonics order 33, for example, a total number of 2688 single measurements with a rotating source are needed. The minimal length of one single measurement is chosen to capture a sufficient amount of the room decay. Even with advanced measurement techniques such as the multiple exponential sweeps method [56], the complete measurement time lasts at least two hours.

Chu [57] made first test measurements with changing temperatures and investigated the influence on the steady-state sound pressure levels and reverberation time. For a temperature change of $0.2\,°\text{C}$, large variations of high-frequency reverberation times were found. During the temperature drift in a model room, a number of 160 continuous averages were made. Due to bad signal-to-noise conditions, Chu had to average 10 sequences of MLS signals for a "quasi-steady-state impulse response" [57]. It is not further investigated how good the steady-state assumption is, that is used as reference for the comparison. As will be shown later, the dependence of the error on the temperature difference is not monotonic. Therefore, an error-free reference measurement is of fundamental importance.

Vorländer and Kob [58] described this problem theoretically. The global temperature change in the room induces a change in the speed of sound. This, in turn, causes a stretching or compression of the room impulse response. For simplification, it is assumed that during the measurement the acoustic system has two different states caused by two temperatures Θ_1 and Θ_2. The second impulse response $p_2(t)$ is then a scaled version of the first impulse response $p_1(t)$ if the remaining conditions stay constant:

$$p_2(t) = p_1(K \cdot t) = p_1 \left(\frac{c(\Theta_2)}{c(\Theta_1)} t \right) \; . \tag{5.5}$$

The time difference between both impulse responses increases with the time in the impulse response. For narrow band signals with the frequency f, both impulse responses add up constructively for multiple time delays of the cycle duration $T = 1/f$. For uneven-numbered multiples of the half cycle period, both impulse responses erase each other due to destructive addition in the resulting averaged response $p_a(t)$. This periodic behavior can be expressed as a modulation of the original impulse response $p(t)$ with a cosine:

$$p_a(t) = p(t) \cdot \cos \left(\pi f t \frac{\Delta c}{c} \right) \; . \tag{5.6}$$

This model of measuring two (or more) impulse responses at different temperatures that are correct on their own and the temperature effect rises then by averaging them out afterwards is a valid assumption for equating sweep measurements. For short sweeps, each frequency is excited only for a short period so that each frequency band can be considered as correct. However, the time to the next average sequence is significantly longer and thus the temperature change is larger. For MLS measurements all frequencies are excited during the complete measurement period, the averaging occurs therefore continuously over the whole measurement time. However, Vorländer and Kob [58] showed that an average over N impulse responses can also be approximated by an overlap of two modulation functions:

$$p_a(t) = p(t)\frac{1}{N}\sum_{n=1}^{N}\cos\left(2\pi ft\left(1 + n\frac{\Delta c}{c}\right)\right) \tag{5.7}$$

$$\approx p(t)\frac{c}{2\pi ftN\Delta c}\left[\sin\left(2\pi ft\left(1 + N\frac{\Delta c}{c}\right)\right) - \sin\left(2\pi ft\right)\right] . \tag{5.8}$$

Vorländer and Kob [58] suggested that the time at which the first destructive overlay occurs is at least as high as twice the reverberation time. Based on this the maximum temperature deviation can be derived as:

$$\Delta\Theta \leq \frac{300°C}{fT} . \tag{5.9}$$

In contrast to Chu [57] and Vorländer and Kob [58], this study focuses on the nowadays more commonly used sweeps as excitations signals. In comparison to Vorländer and Kob [58], who used simulations, or Chu [57], who used a model room, in this study a real size auditorium is used. This allows a realistic study of all relevant effects, such as for example, frequency dependent air attenuation. In addition, local temperature inhomogeneity in large rooms is expected to cause effects on separate reflections in the RIR. The changes of the parameters will have a random character. The realistic experiments will show how large these random components are.

5.3.1. Measurement Setup

The measurements have been conducted in the General Assembly Hall of RWTH University Aachen (for detailed information see Appendix A). As usual, the sound source has been placed on the stage and the microphones have been spread over the audience area. The high quality dodecahedron loudspeaker is able to provide high sound pressure levels with acceptable magnitudes of the nonlinearities. For further details on the measurement equipment, see

Section 2.1.2. Due to high output levels, short sweeps with the length of only six seconds can be used, while preserving a sufficient high signal to noise ratio. This way it can be assumed that every single measurement is free from time variance effects and can be used as reference value. A correct reference value is of major importance, since it is the only basis to evaluate the investigated effects. If the reference value is distorted by the same effect, the difference between both will be smaller or, in case of not monotonic effects, even larger than they actually are.

The air conditioning system has been programmed to switch on and off in periods of five to seven hours. Figure 5.22 (top) shows the measured temperature profile. The first hours after the last persons have left the room are ignored for the measurement. In the first phase, the air conditioning system was deactivated. The air temperature outside was around 27 °C, which results in a relatively, high temperature of 22 °C in the hall. The measurements started in the late afternoon as the outside temperature was slowly decreasing which, explains the slow temperature drift in the first phase. Around 14 hours after the measurement session started, the second phase began and the air conditioning system was activated. The air conditioning system has been in cooling mode since the target temperature was set to 20 °C. After seven hours the air conditioning system was switched off and the temperature increased again.

Figure 5.22 (bottom) shows the change of the temperature during the measurement session in °C per hour. To eliminate the quantizing noise of the temperature sensors, a moving average window is applied on the temperature data. In Figure 5.22 (top), the original and the smoothed temperature profiles are shown together. It can be seen that the measurement noise of the temperature sensors is eliminated while the temperature profile is preserved. The temperature change for the active ventilation system is up to 0.8 °C / h and after the deactivation even more than 1.2 °C / h.[1] The temperature profile after a change of the air conditioning status shows a behavior of an exponential function with negative argument. The temperature gradient is greatest directly after the change of the condition and decreases from this point.

---

[1] For the unprocessed temperature values, the temperature change between a few measurements can even reach up to 2 °C / h.

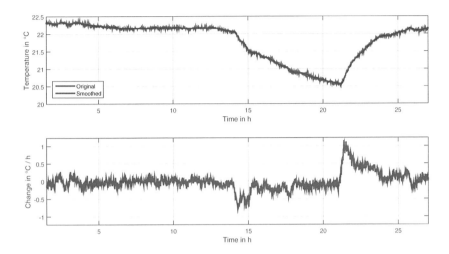

Figure 5.22.: Temperature profile of measurement session (top) and temperature change (bottom) in °C per hour. To eliminate quantizing noise from the temperature sensors, a moving average window is applied. The upper part of the figure shows that the basic temperature profile is preserved.

However, this scenario is realistic, since room acoustic measurements are often conducted directly after or before an event in the auditorium. It is very likely that the air conditioning system is switched off or turned on during the measurement session.

5.3.2. Analysis

The averaging of the impulse responses with different temperatures is done in the post-processing. This procedure has the advantage that the original correct measurements are known, so as to compare the results and give the ability to specify the temperature difference of both measurements freely from the available measurements.

For the following examples, a reference impulse response is chosen that has been measured some time after the air conditioning system has been switched off again (21.5 h). This impulse response is averaged with a subsequent one. Depending on the time difference between the measurement times of both room impulse responses (RIR) and the temperature profile, different temperature changes can be simulated and the effects can be analyzed in detail. The reference impulse responses has a reverberation time of $T_{20} = 1.8$ s in the 1 kHz octave band. According to Eq. (5.9) the maximal temperature change should not exceed

$$\Delta\Theta \leq \frac{300°C}{fT} = \frac{300°C}{1\,\text{kHz} \cdot 1.8\,\text{s}} = 0.167°C .\tag{5.10}$$

For the first example shown in Figure 5.23, the temperature difference is 0.08 °C and therefore smaller than the calculated limit by a factor of 2. The averaged impulse response is shown in red. The reference impulse response is plotted in blue to illustrate the differences. It can be seen that the difference in the amplitudes of both RIRs increases clearly with time in the impulse response. The energy decay curve (EDC) of the averaged impulse response is plotted in green. The dynamic range that is used to evaluate T_{20} is highlighted in dark and bold to visualize the relevant part. The evaluated reverberation time of the averaged RIR is 0.2 s shorter than T_{20} of the reference RIR. This error with a deviation of 10.3 % is very large even considered under perceptual aspects, where the just noticeable difference is 5%. For more sensitive applications such as superpositioning approaches, this error is even less acceptable. The time elapsed between both averaged measurements is only 4.3 minutes. The same temperature differences can also be found in the measurement session for RIR that are taken with a time difference of only 2.2 minutes. This represents the upper range of what is possible in real measurements for high noise levels and large halls.

For the second example, two measurements are used that are acquired with a time separation of 17.8 minutes. The temperature difference is 0.32 °C. The amplitude differences between reference and averaged version exceed 10 dB around 0.8 s. The EDCs of reference and averaged RIR differ significantly. The evaluated reverberation time of 1.2 s disagrees strongly with the reference T_{20} of 1.8 s. The large difference of more than 31.7 % makes this result unusable.

For the third example, the total destructive interference occurs in the middle of the evaluation range of T_{20}. This leads to a sagging EDC in the early part of the evaluation range and a more flat EDC behind that point. In contrast to the previous examples with underestimated reverberation times, the parameter of this example is overestimated (by 32.6 % regarding the reference RIR). The temperature difference is $0.59\,^\circ$C and the time difference between the measurements is 39.1 minutes.

The fourth example shows a quite large temperature difference of $1.12\,^\circ$C, acquired during time interval of nearly 120 minutes. It can be seen that the first completely destructive interference occurs at the beginning of the evaluation range and the second one, located in the evaluation range, is less pronounced. The difference in the evaluated reverberation time is relatively small ($0.03\,$s or 1.4 %) compared to the errors with a smaller temperature difference.

The examples show that the error in the evaluated reverberation time of the averaged impulse response depends on the temperature difference in a complex way. To investigate this relationship more closely, a series of RIRs are averaged in subsequent processing and the reverberation time T_{20} is evaluated. Similar to the four examples shown before, one reference RIR is selected and averaged with a following RIR. For this analysis, the subsequent 500 RIRs are used, leading to 500 pairs of temperature change and reverberation error. This analysis is repeated for 20 reference impulse responses, resulting in a set of around 10,000 pairs of values. Figure 5.24 shows the relative error of T_{20} (regarding the reference RIRs) as function of the temperature difference for the 1 kHz octave band.

It can be seen that the averaged reverberation time is underestimated for temperature differences from $0\,^\circ$C to $0.4\,^\circ$C. The maximum error occurs between $0.25\,^\circ$C and $0.35\,^\circ$C and reaches up to -35%. Above $0.4\,^\circ$C the resulting reverberation time is overestimated for up to more than $+30\%$ at around $0.6\,^\circ$C. From this point the error decreases again for increasing temperatures. The data set only covers a temperature difference of up to $1.5\,^\circ$C; the behavior beyond that point is therefore unknown.

For this example the errors exceed 5% for temperature differences of $0.048\,^\circ$C and 10% for $0.067\,^\circ$C.

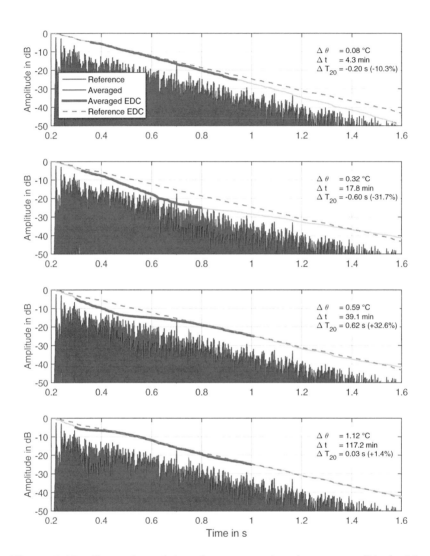

Figure 5.23.: Comparison of the reference room impulse response (blue) with the averaged version (red) for four examples (same microphone position and 1 kHz octave band). For illustration the resulting energy decay curve (EDC) of the averaged version is plotted above the impulse response. The evaluated range of T_{20} is highlighted as bold line.

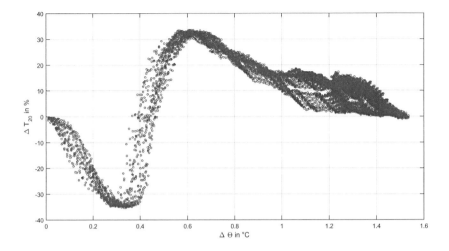

Figure 5.24.: Relative error in T_{20} of averaged impulse response as function of temperature difference during both measurements. The error is referred to the average of the separately evaluated reverberation times.

Frequency Dependence

The time stretching or compressing of the impulse response due to temperature affects the frequency bands in a different way. It is assumed that the basic shape of the function is the same, while the dependence on the temperature change is scaled with frequency. Figure 5.25 shows the relative error of the reverberation time T_{20} for four octave band frequencies from 250 Hz to 2 kHz and the same microphone position. The scaling of the temperature dependence is clearly visible. For the selected examples, the temperature differences where the first error exceeds 5% is scaled nearly exactly with the inverse frequency:

Frequency band	250 Hz	500 Hz	1 kHz	2 kHz
Limit for 5% error	0.236 °C	0.1 °C	0.048 °C	0.019 °C

The temperatures for the maximum positive and negative deviations scale in a similar manner. The relative magnitude of the error is also independent of the

119

considered frequency band and reaches up to 40%. The error does not exceed a maximum value. This results in the worst case scenario: If the lower limit of the evaluation range is congruent with the first zero crossing of the modulation cosine term, the resulting decay is the steepest.

For the $2\,\mathrm{kHz}$ octave band it is also visible that for temperature differences $\geq 0.6\,^{\circ}\mathrm{C}$ the error does not exceed 5%. The larger the temperature difference, the more periods of the modulation cosine term of Eq. (5.6) fall into the evaluation range. Due to an average over several periods, the cosine introduced gradient vanishes and the original decay of the room is revealed. Additionally, less distinct notches occur and the interferences smear, as seen in Figure 5.23 bottom. This fact causes the evaluated reverberation time to fluctuate randomly without a bias into one direction.

Evaluation Range

The influence of the dynamic range on the error behavior is described in Figure 5.26. The $1\,\mathrm{kHz}$ octave band is analyzed for one selected microphone position and the evaluated reverberation times T_{10}, T_{20}, T_{30} and T_{40} are plotted as functions of the temperature difference. For the smallest dynamic range of $10\,\mathrm{dB}$ for T_{10}, the magnitude of the error is the largest and reaches up to 80%. This can be explained by the fact that the evaluation range covers only a small part of the impulse response. If this evaluation range and the zero crossing of the modulation cosine term are congruent, the resulting decay deviates significantly from the original decay. Similar to the already observed behavior, the more periods of the modulation cosine term of Eq. (5.6) appear in the evaluation range the smaller the error. Therefore, the maximum error decreases for larger evaluation ranges. A second effect that can be observed is a scaling of the function on the temperature axis. The higher the dynamic range, the more compressed the curve. For higher dynamic ranges later parts of the room impulse response are analyzed and these are affected even for smaller temperature changes.

A general statement about which dynamic range should be used if temperature changes are expected is not possible. For larger dynamic ranges the magnitude

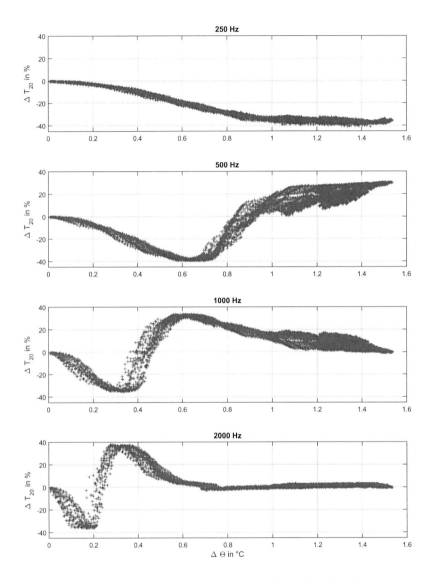

Figure 5.25.: Relative error in reverberation time T_{20} as function of temperature difference for four different octave bands from $250\,\text{Hz}$ to $2\,\text{kHz}$. The shape of the function is scaled by the inverse frequency.

of the error is smaller, but the errors occur already for smaller temperature differences. Whereas, for smaller dynamic ranges the errors occur for higher temperature differences, but the magnitudes are significantly larger.

By defining a certain error limit (i.e. 5%), it can be seen that smaller evaluation ranges exceed this limit for lower temperature differences. Therefore, smaller evaluation ranges should be preferred if temperature changes are expected. Additionally, it has to be taken into account that larger dynamic ranges require lower PSNR values. These can be realized with shorter measurement signals or lower averages, and shorter measurement time implies smaller temperature changes.

Clarity Index

As shown in Eq. (2.30), the clarity index C_{80} can be calculated by using two points of the energy decay curve at $t = 0\,\mathrm{ms}$ and $t = 80\,\mathrm{ms}$. Similar to the findings of the dynamic range dependence, it is assumed that the occurrence of the errors starts at relatively high temperature differences, because the evaluated parts of EDC are early in the RIR. Furthermore, it is assumed that the error is larger, since only two single points of EDC are evaluated and no averaging effect over a larger EDC range occurs. Figure 5.27 shows the evaluated clarity index C_{80} for the octave bands of 500 Hz (top) and 2 kHz (bottom). It can be seen that for the 500 Hz octave band the error is positive for the considered temperature differences. The error increases slowly with temperature and exceeds 1 dB (which corresponds to the JND of C_{80}) around 0.6 °C. For the 2 kHz octave band, the error function is a compressed version. Except for the sign of the error the shape is similar to the reverberation time behavior. The error increases with temperature difference until it reaches the maximum deviation of 4 dB at around 0.5 °C. For further increasing temperature differences, the error decreases again and becomes negative. The scaling of the error functions agrees well with the ratio of the observed frequencies: The center frequency is higher by the factor of 4 and the mean temperature where the error exceeds 1 dB is around 0.15 °C, which corresponds to a quarter of the 500 Hz value.

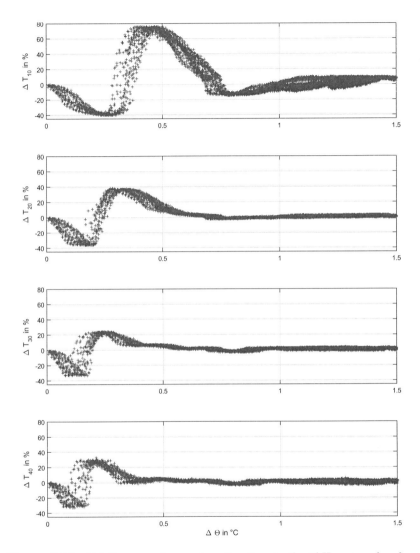

Figure 5.26.: Relative error in reverberation time in the 1 kHz octave band as function of temperature difference for four different evaluation ranges from T_{10} to T_{40}. For higher dynamic ranges, the magnitude of the errors gets smaller but the errors occur already for smaller temperature differences.

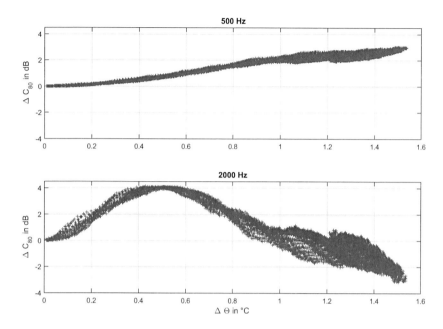

Figure 5.27.: Relative error of clarity index C_{80} in dB as function of temperature difference of both averaged impulse responses. The shape of
the error function scales with the inverse center frequency of the
regarded band.

5.3.3. Summary

The previous analysis showed that the error dependence of the temperature
changes scales inversely with the center frequency of the observed octave band.
To allow a more general description of the empirical error, the number of
influencing variables is decreased. The error is described as a function of the
product of temperature difference and center frequency $\Delta\Theta \cdot f$.

Figure 5.28 shows more than 560.000 temperature differences (averaged in post-
processing) and the resulting error in the reverberation time T_{20} as a scatter plot.
The results include 15 microphone positions and four octave band frequencies.
The mean value of the error is plotted as red solid line and the dotted lines mark

the range including 90% of all data points. It can be seen that the general shape
of the error functions (shown in Figure 5.25) remains the same and confirms
the combination of $\Delta\Theta \cdot f$ to one influence factor. The empirical results of the
measurements are summarized in Table 5.4 for several room acoustic parameters.
For different magnitudes of the factor $\Delta\Theta \cdot f$ the value is listed that marks the
limit where 95 % of the data points fall below that error. For the reverberation
times this value corresponds to the 5th percentile, since the error in reverberation
time is negative for small temperature differences. The clarity index C_{80} shows
an opposite behavior, therefore the 95th percentile is taken.

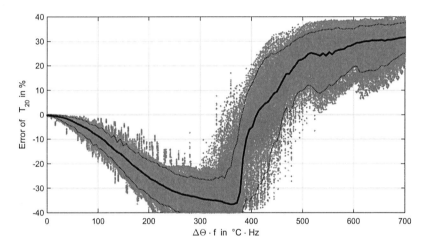

Figure 5.28.: Scatter plot of T_{20} error of of more than 560 000 temperature
differences for several microphone positions and frequency bands.
The solid line marks the median and the dotted lines the 5th and
95th percentile.

The observed effects of single frequency bands and microphone positions in the
previous subsections are confirmed by this wide range summary: The higher
the dynamic range of the evaluated reverberation time the more sensitive the
parameter is to temperature changes. Regarding the maximum factor $\Delta\Theta \cdot f$
that ensures an error below the JND of 5 %, the factor decreases with increasing
dynamic range: starting with 72 for EDT, decreasing over 42 for T_{20} and 30 for
T_{40}. The clarity index C_{80} is less sensitive: The error exceeds the JND of 1 dB

125

for factors $\Delta\Theta \cdot f > 252$.

Although all measurements originate from one room, the reverberation times cover a range from 1.6 s up to 2.1 s due to the variety of frequency bands and microphone positions. For the given range of reverberation times, Table 5.4 provides a good estimation of the error magnitude. However, for clearly deviating reverberation times the results will be different, since a dependence of the error function on the reverberation time is expected.

5.3.4. Conclusion

The temperature change within one impulse response measurement is simulated by averaging two real impulse responses with different temperatures. The observed effects scale exactly with the frequency to which the empirical data of several frequency bands can be combined. Reverberation times with higher evaluation ranges are more sensitive to temperature changes. For T_{20} the product $\Delta\Theta \cdot f$ has to be smaller than 42 to ensure an error $\leq 5\%$. This leads to a maximum tolerable temperature change of $0.08\,°\text{C}$ for $500\,\text{Hz}$ and even $0.021\,°\text{C}$ for $2\,\text{kHz}$. The clarity index C_{80} is more robust to temperature changes. The error in C_{80} exceeds 1 dB for $\Delta\Theta \cdot f \geq 252$. A temperature stability in this range can easily be fulfilled during conventional measurements.

The investigation with real measurements showed that the critical limits for temperature changes are larger than the theoretical predictions from Vorländer and Kob [58]. However, this effect is assumed to be dependent on the reverberation time itself and the empirical results are based only on a limited range of reverberation times (1.6 s to 2.1 s). Therefore, the empirical data can only be used to estimate the error in this specific range. The test measurements can be repeated in different rooms to cover a wider range of reverberation times and to show that the effect is independent of the room.

$f \cdot \Delta\Theta$	EDT	T10	T20	T30	T40	C80
$°C \cdot Hz$	%	%	%	%	%	dB
10	-0.265	-0.457	-0.675	-1.566	-2.104	0.011
20	-0.710	-1.184	-1.534	-2.603	-3.134	0.030
30	-1.456	-2.367	-2.839	-4.019	-4.207	0.057
40	-2.300	-3.500	-4.381	-5.834	-6.789	0.084
50	-2.931	-4.424	-5.457	-7.460	-7.978	0.108
60	-3.790	-6.153	-7.341	-9.417	-10.389	0.146
70	-4.560	-6.676	-8.573	-10.729	-11.904	0.176
80	-5.524	-8.907	-10.300	-13.283	-14.647	0.220
90	-7.202	-11.426	-14.031	-17.387	-18.716	0.310
100	-7.516	-10.979	-14.387	-17.862	-19.009	0.323
110	-7.540	-11.376	-14.009	-17.208	-18.516	0.306
120	-8.678	-13.322	-15.272	-18.525	-20.041	0.330
130	-9.971	-14.502	-17.697	-21.242	-22.957	0.400
140	-10.925	-15.021	-19.438	-21.913	-24.153	0.421
150	-12.109	-17.422	-22.040	-24.673	-26.436	0.487
160	-12.314	-17.267	-22.034	-25.144	-26.805	0.497
170	-14.361	-20.073	-25.529	-28.740	-29.454	0.624
180	-15.087	-20.939	-26.487	-28.914	-29.608	0.633
190	-15.592	-22.446	-27.207	-29.572	-30.516	0.657
200	-16.891	-24.231	-29.112	-30.784	-30.439	0.724
210	-17.910	-25.365	-30.189	-31.259	-30.672	0.777
220	-19.131	-26.924	-31.291	-31.851	-30.634	0.851
230	-19.722	-27.777	-32.236	-32.095	-30.923	0.877
240	-20.297	-28.639	-33.291	-32.432	-30.595	0.915
250	-21.501	-30.920	-34.496	-32.365	-29.655	0.986
260	-22.503	-32.368	-35.415	-32.091	-28.006	1.074
270	-23.704	-33.612	-36.157	-31.556	-25.137	1.141
280	-25.157	-34.997	-36.918	-30.475	-22.374	1.228
290	-26.413	-35.863	-37.207	-27.274	-16.503	1.328
300	-26.952	-36.652	-37.516	-24.400	-11.993	1.332

Table 5.4.: Empirically evaluated error for several room acoustic parameters depending on the product of frequency and temperature difference $f \cdot \Delta\Theta$. Values represent the error size that dropped below 95% of all measurement data.

5.4. Person Scatter

To characterize the room acoustic properties of an auditorium, several positions have to be measured and averaged. ISO 3382-1 suggests a number from 12 to 30 source-receiver combinations, dependent on the size of the room. It is common practice that one person of the measurement staff moves the microphones and loudspeakers while another person starts the measurements. It is unknown how far away the persons has to move away from the microphone to have no influence on the measurement result. In this study, the person is regarded as a static object that introduces shadowing, reflection and scattering effects to the sound field. A special measurement procedure has been designed to indicate the change in the sound field introduced by humans as scattering objects. Different distances and positions of the scattering object are investigated to recognize possible dependencies. Based on this knowledge, recommendations can be derived that prescribe a minimal distance between persons and microphones during room acoustic measurements.

The question of the impact of present objects is not exclusively interesting for room acoustic measurement. In room acoustic simulations, an important question is up to which level of detail rooms have to be modeled to achieve a sufficiently precise result. The analysis how humans influence the measurements can be transferred to parts of the room or contents in the room of similar size, such as pillars or furniture.

The results presented in this section are partly based on results published in Guski and Vorländer [59].

5.4.1. Measurement Setup

The measurements have been conducted in the General Assembly Hall of RWTH Aachen University (for detailed information see Appendix A). The dodecahedron loudspeaker has been placed on the stage at the position of the soloist. A life-sized dummy of an adult person has been used as scattering object. The dummy has been clothed to keep the measurement as realistic as possible (see Figure 5.29 left).

The additional absorption that is introduced by the dummy is $\leq 2\,\mathrm{m}^2$ [51]. This is only a small contribution compared to the equivalent absorption area between $4600\,\mathrm{m}^2$ and $5800\,\mathrm{m}^2$ (depending on the frequency band). The theoretical change in the reverberation time caused by the additional absorption area is less than $1\,\mathrm{ms}$.

The scattering dummy has been placed in the middle of the audience area and three line arrays of microphones have been installed around it (see Figure 5.29 right). Array A has been positioned behind the scattering object in line with object and loudspeaker, modeling a shadowing effect. This construction simulates the situation that the measurement operator stands in line of sight between source and receiver during the measurement. Six microphones with distances from 0.6 m to 3.4 m from the dummy allow an analysis regarding different microphone-scatterer distances. The second array (Array B) has been installed perpendicular to the source-scatterer axis in line with the scattering object. This scenario represents the situation that the measurement operator stands next to the microphone. In this configuration the scattering object might shadow reflections from the side wall or cause new reflections or scattering. The third array (Array C) has been placed between scattering object and source, simulating the measurement operator standing behind the measurement microphone. In this situation the dummy might change the sound field due to new reflections and shadowing of reflections from the back. All three line arrays are used simultaneously.

The measurements have been performed with and without the scattering object in the middle to analyze the differences in detail. The measurement session consists of seven parts, where between every part the condition if the scattering object is present or not is changed (see Figure 5.30). The first four parts last one hour each and the last three parts half the time. The two different conditions are fragmented on purpose, to decorrelate long term variances of the room or the measurement system (i.e. temperature changes in room or equipment) equally on both conditions. This way long term changes do not have a systematic influence on the analysis.

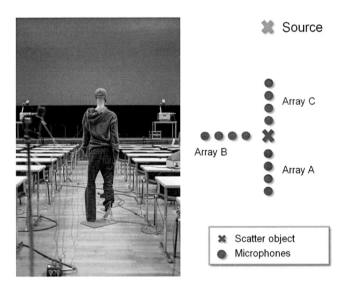

Figure 5.29.: Picture of the person dummy at the center of the microphone arrays (left) and the site of the setup for the scattering measurement session (right).

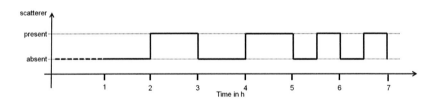

Figure 5.30.: Measurement schedule with the presence of the scattering object.

During the measurements no persons have been present in the auditorium. The measurements with a person inside the auditorium to move the scattering object have been excluded from the analysis. A measurement script started with the acoustic measurement, measured temperature and relative humidity using eight sensors distributed in the room, and saved the results. After a short break of a few seconds, the procedure started all over again, providing around 150 measurements per hour. The large number of repetitions in each measurement part allows a statistical analysis that distinguishes between the scattering object and other factors of random measurement uncertainty. The total number of measurements is around 850.

5.4.2. Analysis

For every single measurement, the room acoustic parameters have been analyzed to study the effect of the scattering object. Figure 5.31 shows examples of the evaluated reverberation time T_{20} of the 500 Hz octave band for different microphone positions. These three examples are selected to present the range of results. Figure 5.31 (top) shows a measurement position where the influence of the scattering object is directly visible. Furthermore, the difference that is introduced by the scattering object is reproducible for each single phase. The presence of the scattering object always causes a decrease of the evaluated reverberation time. For the second example (Figure 5.31 middle), the effect is less clear. Although changes between the phases are visible, the differences are smaller. Figure 5.31 bottom shows an example where no clear effect can be seen. On some changes of the conditions (i.e. after 1 h or after 4 h), no effect on the evaluated parameter is visible, whereas other condition-changes show clear effects.

For the reverberation time this effect is not correlated with the position of the array, the distance to the scattering object, or the frequency band. The magnitude of the relative change in T_{20} is rather small compered to the JND. The reverberation time is evaluated after a decay of -5 dB below the maximum, because its purpose is to estimate the decay of the diffuse sound field and to be independent of direct sound and early reflections. However, a scattering object near the microphone is rather suspected to have an influence on the early part of the impulse response. This explains the small changes of the reverberation time evaluation.

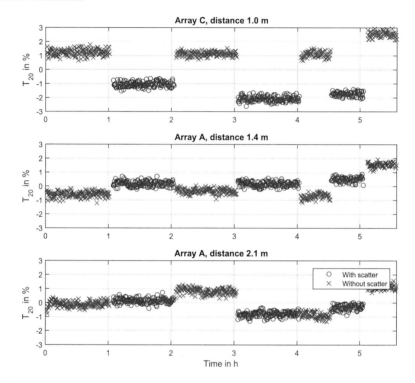

Figure 5.31.: Evaluated reverberation time T_{20} for the 500 Hz octave band for three different microphone positions. These examples show the bandwidth of the results: clear reproducible differences (top), less clear differences (center), and random differences (bottom). The perceivable difference is about 5%.

Clarity Index

The clarity index C_{80} relates early reflections to late reflections and is therefore more sensitive to the scattering object. Figure 5.32 shows the evaluated clarity for microphone Array A and different distances between microphone and person dummy.

For a distance of 0.6 m (Figure 5.32 top), a clear difference between the two scenarios can be seen. Caused by the short distance and the positioning in the line of sight, the differences of C_{80} are quite high (1 dB to 1.5 dB). For a larger distance of 1.4 m, the differences range from 0.6 dB to 1.25 dB, which is slightly smaller but still larger than the JND. For increasing distances the magnitude of the effect decreases up to a distance of 3.1 m, where the difference between the phases is clearly visible and larger than the random variation in the phases, but the differences of around 0.2 dB can be neglected for conventional room acoustic parameters. In the examples of Figure 5.32, it can be observed that the C_{80} values of the scenario without the scattering object stay constant over the complete measurement session. This indicates that no other influences have distorted the measurements.

However, the clear relation between distance and magnitude of error is an exception. Other frequency bands or arrays show results that are more difficult to interpret.

5.4.3. ANOVA

Especially for the reverberation time, no conclusive statement could be made on the influence of the scattering object. To allow an objective and reliable statement about the influence, a statistical analysis is applied. The analysis of variances (ANOVA) compares the mean values and the variances of two groups of data and states if these two groups are independent or if they are random samples from the same population. In this investigation the different microphone positions and the frequency bands are analyzed independent of each other, using a one-way ANOVA. Therefore, all measurement results made in presence of the scattering object are summarized in one group and those with absent object in the second group. Due to the fragmentation of the states of the scattering object over time, the decorrelation between the presence of the scattering object and a long term change in the auditorium or equipment is ensured. The significance level of $p < 0.05$ is defined as the threshold to determine the significance of the scattering object.

133

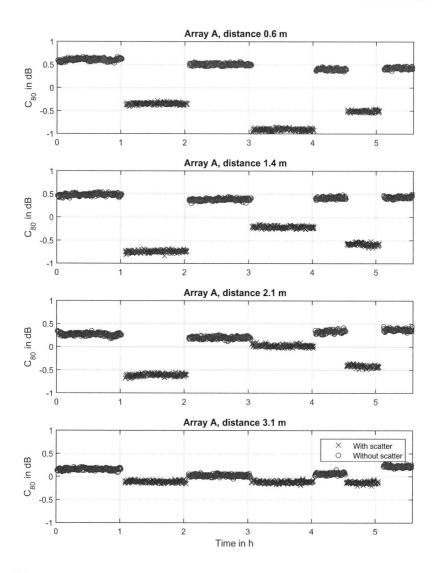

Figure 5.32.: Measured clarity index C_{80} of 1 kHz octave band as function of time for Array A. The differences between measurements with and without scattering object increase for decreasing distances to the scattering object. The values are normalized to the mean of each microphone position. The perceivable difference is about 1 dB.

Reverberation Time

Figure 5.33 shows a summary of 128 one-way ANOVAs (8 frequency bands on x-axis and 16 microphone positions) for the reverberation time T_{20}. For significant differences between both groups ($p < 0.05$), the fields are marked in blue and if no significant relation can be found the results are marked in orange.

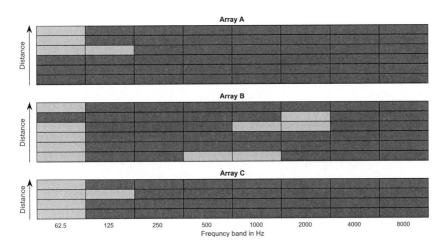

Figure 5.33.: One-way ANOVA analysis for the reverberation time T_{20} of each combination of microphone positions and frequency bands. Significant differences between presence and absence of scattering object are marked in blue (dark) and insignificant results in orange (light). A significance level of $p < 0.05$ is used.

The majority of the analyzed channel-frequency combinations (109 of 128) show a significant influence of the scattering object on T_{20}. Notably the insignificant combinations concentrate at the lowest octave band frequency of 63 Hz. Detailed analysis has shown that this can be explained by increased random fluctuations of the parameter, caused by low signal-to-noise ratios. In the following, the lowest frequency band is therefore not further considered. The remaining insignificant combinations occur rather unsystematically. With exception of the already-discussed lowest band, no correlation with the frequency band can be observed. This behavior is unexpected, since it had been anticipated that the scattering has no impact on the sound field for wavelengths that are larger than the scattering object. For the used person dummy the height corresponds to a frequency of 190 Hz and the width to 700 Hz. Below these frequencies the effect on the measurements should be negligible, but the 125 Hz band almost exclusively shows significant results.

It is assumed that the influence of the scattering object decreases with the distance to the microphone. Considering the results of Figure 5.33, such a relation is not observable. Since the results do not give any rudimentary hint to depend on the microphone position or the frequency band, it is assumed that the few insignificant position-frequency combinations originate from another random source. One possible scenario might be that the scattering object had not been placed at the exact same position. This could cause different deviation of the parameters for two phases with a present scattering object. Since all phases with an object are combined in one group, two phases that are significant for themselves become insignificant if they are combined. However, since only a few combinations might be affected by this circumstance, it is recommended to discontinue this investigation.

Clarity Index

Figure 5.34 shows the ANOVA analysis for the clarity index C_{80}. The majority of the position-frequency combinations show a significant influence of the scattering object. There are even fewer outliers than for the reverberation time analysis. This is in line with the observed larger deviations of C_{80} in the previous section. The insignificant results concentrate at lower frequency bands. The large fluctuations of C_{80} at the 62.5 Hz octave band suggest that a low signal-to-noise ratio is responsible for this effect.

Summary ANOVA

The ANOVA analysis allows for the conclusion that a human sized scattering object in the range of six meters around the microphone in most cases has a significant influence on the room acoustic parameters. However, this reveals

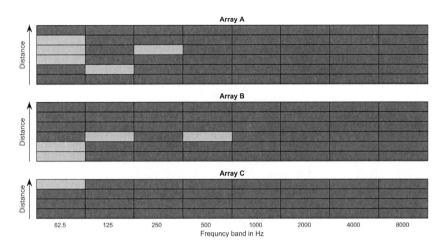

Figure 5.34.: One-way ANOVA analysis of clarity index C_{80}. Significant differences between presence and absence of scattering object are marked in blue (dark) and insignificant results in orange (light). A significance level of $p < 0.05$ is used.

nothing about the magnitude of the error. The remaining unclear behavior occurs in the 62.5 Hz octave band and is therefore located below the normal frequency range (250 Hz - 2 kHz) and even below the extended frequency range (125 Hz - 4 kHz) defined by ISO 3382-1 [3].

5.4.4. Summary

The large amount of data and numerous dependencies complicate the general interpretation of the data. Therefore, in this chapter the data is summarized as well as possible to get an impression of the error sizes and the influence factors. Up to 150 single measurements of each phase are combined and represented by their median value. Further, only the (relative) changes between the phases are observed, separately for each parameter, frequency band, and microphone position.

Figure 5.35 gives an overview of the results. The plots in the left column show the relative change in the reverberation time T_{20} and in the right column the deviation of the clarity index C_{80}. The rows represent the five octave band frequencies from 125 Hz to 2 kHz. Each plot is divided into three blocks of data, showing the three microphone arrays. Inside a block the microphone positions are arranged with increasing distance from left to right. The x-symbols indicate the absence and the circles the presence of the scattering object (c.f. Figure 5.35).

The changes of the reverberation time between the different phases are always smaller than 4%. The majority of phase changes show differences that are even below 2%. No dependence of the frequency band on the magnitude of the error can be observed. Looking at the different microphone arrays, two cases are noticeable: Array A at 1 kHz and 2 kHz. The deviations tend to decrease with increasing distances from the scattering object. However, this cannot be interpreted as a general behavior, since the remaining 13 array-frequency combinations do not show a clear dependence. The comparison of the three microphone arrays, does not reveal differences in the magnitude of the error.

In the right column the absolute differences of the clarity index C_{80} are shown. The influence of the dummy on the parameter increases clearly with frequency. For octave bands > 1 kHz the changes are in the order of the JND of clarity. For lower frequencies the deviations between the phases rarely exceed 0.3 dB.
Array A (representing the scattering object in line of sight between loudspeaker and microphone) shows the largest detected differences in the 1 kHz and 2 kHz band, almost reaching 2 dB for individual cases. For the remaining cases the three microphone arrays have the same behavior.

5.4.5. Conclusion

The investigation of the human-sized scattering object on the measurements showed that a significant influence can be observed. The evaluated room acoustic parameters show a clear change larger than the remaining random variations if the human dummy is present or absent in the audience area. An ANOVA analysis reveals that the majority of the investigated microphone positions and frequency bands show a significant influence on the scattering object. This holds for all investigated positions in the range of 6 m.

The magnitude of the change differs for the considered parameters. For the reverberation time, the changes are below 2% in most cases and are therefore not relevant for standard room acoustic measurements. Furthermore, the reverberation time shows no dependence on the microphone-scatterer distance.

The clarity index C_{80} is more sensitive to the scattering object, since the direct sound and the early reflections, which might be influenced by the object, have a significant impact on the parameter. For frequencies > 500 Hz the error is in the order of JND and therefore becomes critical even for conventional room acoustic measurements.

The same guidelines apply for human-sized objects in room acoustic simulations.

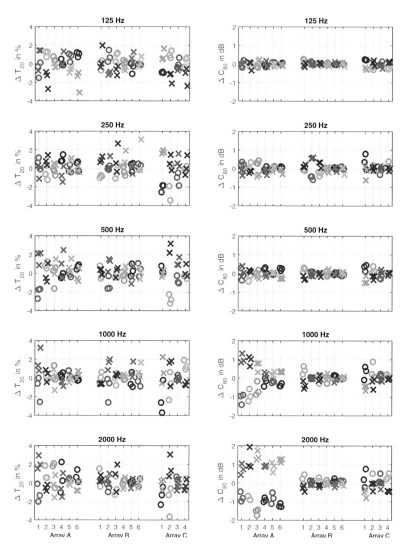

Figure 5.35.: Detailed overview of changes between phases of reverberation time T_{20} (left column) and clarity index C_{80} (right). The three blocks in a plot represent the three microphone arrays. The x-symbols indicate the absence and the circles the presence of the scattering object.

6

Conclusion and Outlook

Interfering external influences on room acoustic measurements and the errors in the determined parameters were investigated in this study. Therefore, six frequently occurring influences were analyzed in detail:

- stationary noise,
- impulsive noise,
- air movement in the auditorium,
- temperature changes between measurements,
- temperature changes within one measurement, and
- human sized scatter objects.

Special measurement setups were designed that control one influence factor while other influence factors were kept as constant as possible. The relationship between the size of the influence factor and the evaluated room acoustic parameters were investigated. The resulting dependencies of the measurements were compared with theoretical models of the interfering effect. The determined empirical results have been documented to allow an easy estimation of the error size and a comparison of different error sources.

Noise inevitably occurs during room acoustic measurements and is one of the largest error sources. Stationary noise and impulsive noise affect measurements in a different way and were therefore handled separately.

For stationary noise, various noise compensation techniques can be applied before evaluating the room acoustic parameters. A theoretical approach that models the envelope of a simple measured room impulse response has been developed to perform simulations for an analysis of different noise compensation techniques. The resulting room acoustic parameters were analyzed numerically subject to the modeled peak signal-to-noise ratio (PSNR). In this study five noise compensation methods were investigated and it was determined that these approaches differ significantly regarding the systematic error. Some methods result in large errors, even for sufficiently high PSNR values, and other methods provide results for obviously insufficient PSNRs. Very large errors have been detected if no post-processing is applied. This thesis demonstrates that it is important to use noise compensation techniques.

To verify the model results, a large number of acoustic measurements have been conducted with changing output levels. In this way, numerous real impulse responses from the same system have been provided that only differ in the contained noise. These measurements clearly validate the developed model approach.

The measurement results were also used to determine the empirical random variations of the parameters depending on the PSNR values. The results are similar for almost all noise compensation techniques. A function has been specified to estimate the maximum error that occurs for the reverberation time T_{20}.

The previously investigated noise compensation methods assume stationary noise conditions, but in reality occasionally impulsive noise appears. Especially for the commonly used sweep measurement technique, the occurrence of impulsive noise during the measurements results in large errors of the impulse response. Since the discussed noise compensation methods are not able to handle these kinds of distortions, large errors of the evaluated room acoustic parameters are the consequence.

In this thesis, a new technique for automatic detection of impulsive noise has been presented. It was shown that impulsive noise events can cause significant errors in the evaluated room acoustic parameters even if the sound pressure level is below the excitation signal. The proposed technique is able to detect these disruptions, whereas a visual inspection is often impossible. This method can be applied to the impulse response directly after the measurement to allow for an immediate repetition of the measurement if necessary.

In further studies the proposed impulsive noise detection can be used to develop an impulsive noise compensation algorithm. A band pass filter can be applied at the time of the occurrence of the impulsive noise to reduce the impact and just let pass the instantaneous frequency of the sweep including a safety band below for the decaying frequencies. A spectral subtraction based on interpolated levels of excitation and impulsive noise can be used to enhance the correction. Nonetheless, this is a serious manipulation of the measurement and it has to be very reliable to apply.

The second part of this thesis investigated the violation of the time-invariant systems assumption. Air movements in auditoria cause changes in the room acoustic conditions. In this study a measurement session has been designed to investigate the impact on the evaluated room acoustic parameters. It has been demonstrated that the ventilation system increases the random variances of the parameters significantly. The size of the effect increases for higher frequencies.

Variations of temperature and relative humidity were analyzed separately for changes during and between measurements. The meteorological conditions in an auditorium have a direct influence on the room acoustic properties. It has been investigated how large a difference in temperature and humidity is acceptable to ensure that two measurements made under different conditions can be compared. The diffuse field theory is suitable to predict and correct the temperature influence. However, for some measurements a second effect occurred that was significantly larger in magnitude. The size of this error also scales with the temperature difference, but the sign of this error was random for different microphone positions or frequency bands. Therefore, a prediction or compensation is not possible. The size of the effect has been documented to enable an estimation of the influence. The permitted temperature changes to ensure a maximum error size have been reported to ensure the comparability of two measurements.

Further measurements should be applied to investigate the second effect in detail and determine possible influence factors to enable a prediction or at least a more general estimation of the error size. These measurements should also capture the vertical temperature profile to discover possible relations.

Temperature changes during a measurement, on the other hand, lead to destructive interferences and therefore large errors in one measurement. This thesis demonstrated with real measurements that the error is related to the temperature difference in a complex way and scales clearly with frequency. It has been shown that the sensitivity of the reverberation time is clearly higher than the theoretical predictions.

Persons as scattering objects in the auditorium were also analyzed as influence factor. Due to additional reflections, scattering, or shadowing of reflections from the object, the temporal structure of the impulse response and the evaluated parameters might change. A measurement setup was designed with a life-sized human dummy in the middle of a microphone array. The statistical analysis of variances showed that there is a significant difference between measurements with and without the scattering object in the area of 6 m around the object. Despite the statistical significance, the magnitudes of change in reverberation time are rather small and can therefore be neglected for standard room acoustic measurements. The clarity index is characterized by the temporal structure of the RIR and is therefore more sensitive to scattering objects. A minimal distance of 10 m between person and microphone should be kept to guarantee precise measurements.

These findings can also be applied to room acoustic simulations: For reverberation time simulations, human-sized objects (such as pillars or furniture) can be neglected without large influence on the simulated parameter. However, for the clarity index these objects should be included in the measurements of the nearby environment of source and receiver.

In conclusion, the results of the study show that noise is one of the largest error sources. It is recommended to use advanced stationary noise compensation techniques to guarantee a robust result. Although, these noise compensation methods also reduce the influence of impulsive noise, it is not possible to compensate this effect completely. Therefore, it is of major importance to apply the proposed impulsive noise detection algorithm and repeat the measurements if impulsive noise was present.

These investigations showed a large influence of inter- and intra-measurement temperature changes. Particularly the reverberation time measurements are sensitive to even minor changes and require a careful monitoring of the conditions. Although the influence of air movement on room acoustic parameters is clear, the magnitudes are small and can be neglected for conventional measurements. The same also applies to the influence of human-sized scatter objects on the reverberation time. The clarity index, however, is more sensitive and requires a minimum distance of 10 m.

The comparison of the error sizes shows that the air conditioning system in the auditoria should be activated, since the room acoustic parameters are less sensitive to the air movements, whereas temperature changes have more of an impact.

The final conclusion is that the measurement standards in room acoustics could be improved in future revisions. Apart from the well-known problem introduced by inaccurate omnidirectionality of the sources used, several steps in the measurement procedure and the post-processing are likely to be included in more narrow specifications. This applies in particular to the usage of stationary noise compensation techniques and the awareness of the sensitivity of sweeps to impulsive noise. Furthermore the standard should include guidelines the minimal distance between persons and microphones and temperature changes during single measurements.

Danksagung

Das Ende dieser Promotion ist eine gute Gelegenheit mich bei den vielen Menschen zu bedanken, die mich dabei unterstützt und das alles somit ermöglicht haben.

Prof. Michael Vorländer möchte ich danken für die Möglichkeit an seinem Institut arbeiten und promovieren zu dürfen und für die Betreuung während dieser Zeit. Mein Dank gilt auch Prof. Heberling für die Übernahme des Zweitgutachtens und für die konstruktiven Hinweise.

Ganz besonders möchte ich mich bei allen Mitarbeitern des Instituts für Technische Akustik für die sehr freundliche und hilfsbereite Arbeitsatmosphäre bedanken. An dieser Stelle möchte ich stellvertretend einige Personen besonders hervorheben: Dr.-Ing. Gottfried Behler, Dr.-Ing. Pascal Dietrich, Ingo B. Witew und Dr.-Ing. Markus Müller-Trapet möchte ich besonders danken, da sie mir mit ihrem Fachwissen zu Raumakustik, Messtechnik und Signalverarbeitung besonders in der Anfangszeit immer zur Seite standen. Den Kollegen aus dem Developer-Team der ITA-Toolbox danke ich für die Entwicklung eines mächtigen Werkzeugs, welches mir tagtäglich viele Aufgaben deutlich vereinfacht hat.

Für die zahlreichen fachlichen Diskussionen, aber auch die motivierenden und unterhaltsamen Mittags- und Kaffeepausen möchte ich mich besonders bei Josefa Oberem, Ramona Bomhardt, Prof. Janina Fels, Marcia Lins, Rob Opdam, Jan Gerrit-Richter und Dr.-Ing. Marc Aretz bedanken.

Des Weiteren bedanke ich mich bei Karin Charlier für die Unterstützung bei den vielen bürokratischen Angelegenheiten. Der elektronischen und der mechanischen Werkstatt gilt mein Dank für die unkomplizierte und konstruktive Hilfe bei zahlreichen Projekten.

Ganz besonderer Dank gilt meiner Frau Valeria. Seit dem Studium ist sie immer an meiner Seite und hat mich zu jeder Zeit auf meinem Weg unterstützt. Danken möchte ich auch meiner Familie die mich stets ermutigt hat. Mein ganz besonderer Dank gilt natürlich meinen Eltern, ohne die das alles nicht möglich gewesen wäre. Diese Arbeit ist meinen Eltern gewidmet.

Bibliography

[1] H. Wallach, E. B. Newman, and M. R. Rosenzweig. "A Precedence Effect in Sound Localization". In: *The Journal of the Acoustical Society of America* 21.4 (1949), pp. 468–468.

[2] H. Haas. "Über den Einfluß eines Einfachechos auf die Hörsamkeit von Sprache". In: *Acta Acustica united with Acustica* 1.2 (1951), pp. 49–58.

[3] ISO 3382-1. "Acoustics – Measurement of room acoustic parameters – Part 1: Performance spaces". In: *International Organization for Standardization* (2009).

[4] X. Pelorson, J.-P. Vian, and J.-D. Polack. "On the variability of room acoustical parameters: Reproducibility and statistical validity". In: *Applied Acoustics* 37.3 (1992), pp. 175 –198.

[5] A. Lundeby, T. E. Vigran, H. Bietz, and M. Vorländer. "Uncertainties of Measurements in Room Acoustics". In: *Acta Acustica united with Acustica* 81.4 (1995), pp. 344–355.

[6] J. S. Bradley. "An International Comparison of Room Acoustics Measurement Systems". In: *Internal Report, Institute for Research in Construction, National Research Council Canada* 1 (1996), pp. 1–130.

[7] B. F. G. Katz. "International Round Robin on Room Acoustical Impulse Response Analysis Software 2004". In: *Acoustics Research Letters Online* 5.4 (2004), pp. 158–164.

[8] A. Oppenheim, R. Schafer, and J. Buck. *Discrete-time signal processing.* Prentice-Hall signal processing series. Prentice Hall, 1999.

[9]   J. Ohm and H. Lüke. *Signalübertragung: Grundlagen der digitalen und analogen Nachrichtenübertragungssysteme*. Springer-Lehrbuch. Springer, 2010.

[10] M. Vorländer. *Auralization*. Springer-Verlag, Berlin, 2007.

[11] Institute of Technical Acoustics, RWTH Aachen University. *ITA-Toolbox: A MATLAB Toolbox for the needs of acousticians*. http://www.ita-toolbox.org/. 2015.

[12] M. Cohn and A. Lempel. "On fast M-sequence transforms (Corresp.)" In: *Information Theory, IEEE Transactions on* 23.1 (Jan. 1977), pp. 135–137.

[13] J. Vanderkooy. "Aspects of MLS Measuring Systems". In: *J. Audio Eng. Soc* 42.4 (1994), pp. 219–231.

[14] S. Müller and P. Massarani. "DISTORTION IMMUNITY IN IMPULSE RESPONSE MEASUREMENTS WITH SWEEPS". In: *International Congress on Sound & Vibration* 18 (2011).

[15] G.-B. Stan, J.-J. Embrechts, and D. Archambeau. "Comparison of Different Impulse Response Measurement Techniques". In: *J. Audio Eng. Soc* 50.4 (2002), pp. 249–262.

[16] S. Müller and P. Massarani. "Transfer-Function Measurement with Sweeps". In: *J. Audio Eng. Soc* 49.6 (2001), pp. 443–471.

[17] J. L. Svensson Peter; Nielsen. "Errors in MLS Measurements Caused by Time Variance in Acoustic Systems". In: *J. Audio Eng. Soc* 47.11 (1999), pp. 907–927.

[18] O. Kirkeby and P. A. Nelson. "Digital Filter Design for Inversion Problems in Sound Reproduction". In: *J. Audio Eng. Soc* 47.7/8 (1999), pp. 583–595.

[19] A. Farina. "Simultaneous measurement of impulse response and distortion withaswept-sinetechnique". In: 2000, pp. 18–22.

[20] D. G. Ciric, M. Markovic, M. Mijic, and D. Sumarac-Pavlovic. "On the effects of nonlinearities in room impulse response measurements with exponential sweeps". In: *Applied Acoustics* 74.3 (2013), pp. 375 –382.

[21] H. Kuttruff. *Acoustics ? An Introduction*. Taylor & Francis, New York, 2007, pp. 86–88.

[22] H. E. Bass, L. C. Sutherland, A. J. Zuckerwar, D. T. Blackstock, and D. M. Hester. "Atmospheric absorption of sound: Further developments". In: *The Journal of the Acoustical Society of America* 97.1 (1995), pp. 680–683.

[23] ISO 9613-1. "Acoustics -Attenuation of sound during propagation outdoors - Part 1: Calculation of the absorption of sound by the atmosphere". In: *International Organization for Standardization* (1993).

[24] ISO 18041. "Acoustic quality in rooms - Specifications and instructions for the room acoustic design". In: *International Organization for Standardization* (2015).

[25] M. R. Schroeder. "New Method of Measuring Reverberation Time". In: *Acoustical Society of America Journal* 37 (1965), pp. 409–412.

[26] V. Jordan. "Room acoustics and architectural acoustics development in recent years". In: *Applied Acoustics* 2.1 (1969), pp. 59 –81.

[27] ISO 354. "Acoustics – Measurement of sound absorption in a reverberation room". In: *International Organization for Standardization* (2003).

[28] ISO 17497-1. "Acoustics - Sound-scattering properties of surfaces - Part 1: Measurement of the random-incidence scattering coefficient in a reverberation room". In: *International Organization for Standardization* (2006).

[29] ISO 3741. "Acoustics - Determination of sound power levels and sound energy levels of noise sources using sound pressure - Precision methods for reverberation test rooms". In: *International Organization for Standardization* (2013).

[30] ISO 140. "Acoustics - Measurement of sound insulation in buildings and of building elements". In: *International Organization for Standardization* (1998).

[31] H. Seraphim. "Über die Wahrnehmbarkeit mehrerer Rückwürfe von Sprachschall". In: *Acta Acustica united with Acustica* 11.2 (1961-01-01T00:00:00), pp. 80–91.

[32] R. Thiele. "Richtungsverteilung und Zeitfolge der Schallrückwürfe in Räumen". In: *Acta Acustica united with Acustica* 3.4 (1953), pp. 291–302.

[33] W. Reichardt, O. Alim, and W. Schmidt. "Abhaengigkeit der Grenzen zwischen brauchbarer und unbrauchbarer Durchsichtigkeit von der art des Musikmotives, der Nachhallzeit und der Nachhalleinsatzzeit". In: *Applied Acoustics* 7.4 (1974), pp. 243 –264.

[34] H.-P. Seraphim. "Untersuchung über die Unterschiedsschwellen exponentiellen Abklingens von Rauschbandimpulsen". In: *Acustica* 8 (1958), pp. 280–284.

[35] T. J. Cox, W. J. Davies, and Y. W. Lam. "The Sensitivity of Listeners to Early Sound Field Changes in Auditoria". In: *Acta Acustica united with Acustica* 79.1 (1993), pp. 27–41.

[36] J. Bradley, R. Reich, and S. Norcross. "A just noticeable difference in C50 for speech". In: *Applied Acoustics* 58.2 (1999), pp. 99 –108.

[37] M. Ahearn, M. Schaeffler, M. Vigeant, and R. D. Celmer. *The Just Noticeable Difference in the Clarity Index for Music, C80*. Tech. rep. University of Hartford Acoustics, 2009.

[38] J. Klein, M. Pollow, P. Dietrich, and M. Vorländer. "Room Impulse Response Measurements with Arbitrary Source Directivity". In: (2013). 1 CD-ROM, pp. 1–4.

[39] M. Guski and M. Vorländer. "Comparison of Noise Compensation Methods for Room Acoustic Impulse Response Evaluations". In: *Acta Acustica united with Acustica* 100.2 (2014), pp. 320–327.

[40] R. Kürer and U. Kurze. "Integrationsverfahren zur Nachhallzeitauswertung". In: *Acustica* 19 (1967), pp. 313–322.

[41] L. Faiget, C. Legros, and R. Ruiz. "Optimization of the impulse response length: Application to noisy and highly reverberant rooms". In: *Journal-Audio Engineering Society* 46 (1998), pp. 741–750.

[42] W. T. Chu. "Comparison of reverberation measurements using Schroeder's impulse method and decay-curve averaging method". In: *Journal of the Acoustical Society of America* 63.5 (1978), pp. 1444–1450.

[43]   C. Hak, R. Wenmaekers, and L. C. J. van Luxemburg. "Measuring Room Impulse Responses: Impact of the Decay Range on Derived Room Acoustic Parameters". In: *Acta Acustica united with Acustica* 98.6 (2012), pp. 907–915.

[44]   M. Karjalainen, P. Antsalo, A. Makivirta, T. Peltonen, and V. Valimaki. "Estimation of modal decay parameters from noisy response measurements". In: *J. Audio Eng. Soc* 50.11 (2002), pp. 867–878.

[45]   N. Xiang. "Evaluation of reverberation times using a nonlinear regression approach". In: *The Journal of the Acoustical Society of America* 98.4 (1995), pp. 2112–2121.

[46]   C. Huszty and S. Sakamoto. "Application of calculating the reverberation time from room impulse responses without using regression". In: *Forum Acusticum 2011* (2001), pp. 1929–1934.

[47]   J. Bradley. "Review of objective room acoustics measures and future needs". In: *Applied Acoustics* 72.10 (2011), pp. 713 –720.

[48]   M. Guski and M. Vorländer. "Impulsive noise detection in sweep measurements". In: *Acta Acustica united with Acustica* 101.4 (2015), pp. 723–730.

[49]   S. Müller. *Measuring transfer-functions and impulse responses. Handbook of signal processing in acoustics.* Springer New York, 2009.

[50]   A. Farina. "Advancements in Impulse Response Measurements by Sine Sweeps". In: *Audio Engineering Society Convention 122*. May 2007.

[51]   H. Kuttruff. *Room Acoustics.* Taylor & Francis, 4th edition, 2000.

[52]   M. Pollow, P. Dietrich, and M. Vorländer. "Room Impulse Responses of Rectangular Rooms for Sources and Receivers of Arbitrary Directivity". In: (2013). 1 CD-ROM, pp. 1–4.

[53]   A. Novak, L. Simon, F. Kadlec, and P. Lotton. "Nonlinear System Identification Using Exponential Swept-Sine Signal". In: *Instrumentation and Measurement, IEEE Transactions on* 59.8 (Aug. 2010), pp. 2220–2229.

[54] M. Guski and M. Vorländer. "Uncertainty of room acoustic parameters caused by air movement and temperature changes". In: *DAGA 2014 40. Jahrestagung für Akustik 10.-13. März 2014 in Oldenburg* 40 (2014), pp. 439–440.

[55] ISO GUIDE 98-3. "Uncertainty of measurement - Part 3: Guide to the expression of uncertainty in measurement (GUM:1995)". In: *International Organization for Standardization* (1995).

[56] P. Dietrich, B. Masiero, and M. Vorländer. "On the Optimization of the Multiple Exponential Sweep Method". In: *J. Audio Eng. Soc* 61.3 (2013), pp. 113–124.

[57] W. T. Chu. "Time-variance Effect on the Application of the M-Sequence Correlation Method for Room Acoustical Measurements". In: *Proceedings of the 15th International Congress on Acoustics (ICA 95)* IV (1995), pp. 25–28.

[58] M. Vorländer and M. Kob. "Practical aspects of MLS measurements in building acoustics". In: *Applied Acoustics* 52.3â??4 (1997). - comparison old mhod vs MLS -, pp. 239 –258.

[59] M. Guski and M. Vorländer. "The influence of single scattering objects for room acoustic measurements". In: *DAGA 2015* (2015).

A

General Assembly Hall

The General Assembly Hall of RWTH Aachen University (also called "Aula 1") has a rectangular shape and a volume of approximately 5500 m^3. Figure A.1 shows a photograph of the hall from the front row of a balcony. Aula 1 is primarily used for lectures and classical concerts. It has a capacity of 600 seats, whereof about 500 are located in the main audience area on the floor and the remaining seats are placed on balconies around this area.

Figure A.1.: Picture of General Assembly Hall of RWTH Aachen University.

B

Noise Compensation Methods

B.1. Systematic Error

PSNR dB	Method A %	Method B %	Method C %	Method D %	Method E %
30	738.312	-26.419	NaN	-0.000	NaN
35	702.578	-14.275	NaN	-0.000	NaN
40	623.668	-6.324	NaN	-0.000	NaN
45	532.290	-0.925	NaN	-0.000	NaN
50	421.936	1.847	5.112	-0.000	0.001
55	227.259	1.874	3.149	-0.000	0.001
60	13.136	1.027	1.419	-0.000	0.000
65	2.612	0.455	0.573	-0.000	0.000
70	0.749	0.184	0.220	-0.000	0.000
75	0.230	0.071	0.082	-0.000	0.000

Table B.1.: Systematic error of reverberation time T_{30} in percent as function of the peak signal-to-noise ratio (PSNR) and for the five investigated noise compensation methods.

B.2. Random Error

PSNR dB	Method A %	Method B %	Method C %	Method D %	Method E %
40	596.042	-9.971	NaN	-0.000	NaN
45	496.384	-5.683	NaN	-0.000	NaN
50	404.017	-2.488	NaN	-0.000	NaN
55	322.543	0.041	NaN	-0.000	NaN
60	238.459	1.389	3.462	-0.000	0.000
65	111.244	1.248	2.051	-0.000	0.000
70	6.595	0.664	0.908	-0.000	0.000
75	1.422	0.290	0.363	-0.000	0.000
80	0.412	0.117	0.139	-0.000	0.000
85	0.127	0.045	0.052	-0.000	0.000

Table B.2.: Systematic error of reverberation time T_{40} in percent as function of the peak signal-to-noise ratio (PSNR) and for the five investigated noise compensation methods.

PSNR dB	Method A %	Method B %	Method C %	Method D %	Method E %
20	1191.488	-23.063	-5.347	-0.000	NaN
25	1208.661	-1.548	9.594	-0.000	0.008
30	1019.955	4.138	8.798	-0.000	0.001
35	75.052	3.039	4.551	-0.000	0.000
40	13.281	1.493	1.961	-0.000	0.000
45	3.776	0.637	0.782	-0.000	0.000
50	1.159	0.253	0.299	-0.000	0.000
55	0.363	0.097	0.111	-0.000	0.000
60	0.114	0.036	0.040	-0.000	0.000
65	0.036	0.013	0.014	-0.000	0.000

Table B.3.: Systematic error of early decay time EDT in percent as function of the peak signal-to-noise ratio (PSNR) and for the five investigated noise compensation methods.

PSNR dB	Method A dB	Method B dB	Method C dB	Method D dB	Method E dB
11	4.00	-0.15	0.00	0.00	0.00
13	2.52	-0.05	0.04	0.00	0.00
15	1.59	-0.01	0.05	0.00	0.00
17	1.01	0.01	0.05	0.00	0.00
19	0.63	0.02	0.04	0.00	0.00
21	0.40	0.02	0.03	0.00	0.00
23	0.25	0.02	0.03	0.00	0.00
25	0.16	0.01	0.02	0.00	0.00
27	0.10	0.01	0.01	0.00	0.00
29	0.06	0.01	0.01	0.00	0.00

Table B.4.: Systematic error of sound strength G in dB as function of the peak signal-to-noise ratio (PSNR) and for the five investigated noise compensation methods.

PSNR dB	Method A %	Method B %	Method C %	Method D %	Method E %
30	11.721	57.479	NaN	364.853	NaN
35	8.550	16.636	NaN	107.365	NaN
40	7.632	9.890	9.314	48.003	9.938
45	10.654	6.540	6.821	18.946	6.565
50	10.930	5.265	5.595	6.545	5.149
55	32.393	3.431	3.639	3.108	3.406
60	16.761	2.387	2.491	2.059	2.153
65	7.832	0.814	0.826	0.824	0.809
70	0.784	0.472	0.478	0.471	0.468
75	0.476	0.445	0.446	0.446	0.444
80	0.429	0.427	0.427	0.428	0.427
85	0.393	0.394	0.394	0.395	0.395

Table B.5.: Relative random error of the reverberation time T_{30} as function of the PSNR value for the five investigated noise compensation methods. Values are defined empirically by determining the 95th percentile of measurement data.

PSNR dB	Method A %	Method B %	Method C %	Method D %	Method E %
30	12.014	56.354	NaN	496.266	NaN
35	8.495	16.237	NaN	267.574	NaN
40	8.269	10.561	NaN	236.399	NaN
45	11.746	7.254	NaN	177.297	NaN
50	7.060	6.358	8.702	87.162	8.324
55	10.662	4.317	4.544	40.905	4.602
60	10.881	4.352	4.442	4.216	3.942
65	19.171	2.337	2.495	1.828	2.241
70	13.277	1.047	1.046	1.053	0.997
75	7.840	0.572	0.583	0.568	0.542
80	0.618	0.466	0.471	0.455	0.463
85	0.424	0.409	0.409	0.410	0.408

Table B.6.: Relative random error of the reverberation time T_{40} as function of the PSNR value for the five investigated noise compensation methods. Values are defined empirically by determining the 95th percentile of measurement data.

C

Inter-Measurement Temperature Change

$\Delta\Theta$ °C	125 Hz %	250 Hz %	500 Hz %	1000 Hz %	2000 Hz %	4000 Hz %
-0.1	2.689	1.540	1.839	1.703	2.661	3.344
-0.3	2.283	1.338	1.370	1.461	2.119	3.356
-0.5	2.527	1.352	1.744	2.248	2.807	5.229
-0.7	2.875	1.894	2.335	3.564	3.914	8.225
-0.9	3.016	2.305	2.392	3.754	3.382	8.505
-1.1	3.244	2.719	2.401	3.926	3.720	8.163
-1.3	3.631	2.870	2.999	4.551	6.181	10.061
-1.5	3.946	3.266	3.092	5.253	6.181	8.354
-1.7	4.252	3.493	3.357	6.314	6.566	10.224
-1.9	4.964	3.713	3.616	7.722	8.254	10.407
-2.1	5.607	2.802	3.950	8.446	9.368	10.140

Table C.1.: Calculated variance in measurement of EDT based on the meteorological conditions of both measurements. Deviations are listed depending on the temperature difference.

$\Delta\Theta$ °C	125 Hz %	250 Hz %	500 Hz %	1000 Hz %	2000 Hz %	4000 Hz %
-0.1	1.601	3.951	2.244	2.061	2.673	1.339
-0.3	1.532	3.669	1.245	1.789	1.886	1.658
-0.5	1.385	2.517	1.449	1.798	1.845	1.966
-0.7	1.758	3.517	1.972	2.443	2.356	2.474
-0.9	2.273	6.410	2.162	2.555	2.647	3.060
-1.1	2.885	6.385	2.047	3.048	3.269	2.850
-1.3	2.887	5.405	2.338	3.045	3.468	3.710
-1.5	3.227	6.018	2.319	3.348	4.941	4.455
-1.7	3.191	6.377	2.997	3.643	6.051	5.505
-1.9	3.330	6.832	3.566	4.753	7.505	7.672
-2.1	3.355	7.399	3.796	5.659	8.671	8.843

Table C.2.: Calculated variance in measurement of T_{30} based on the meteorological conditions of both measurements. Deviations are listed depending on the temperature difference.

$\Delta\Theta$ °C	125 Hz %	250 Hz %	500 Hz %	1000 Hz %	2000 Hz %	4000 Hz %
-0.1	2.219	2.313	3.489	4.270	5.261	2.906
-0.3	1.801	2.466	3.092	4.125	4.360	2.136
-0.5	1.826	1.882	2.487	3.918	4.521	2.306
-0.7	1.997	2.495	2.861	4.280	4.397	2.599
-0.9	2.490	2.879	3.252	4.747	5.301	3.060
-1.1	3.119	3.338	3.376	5.213	5.633	3.360
-1.3	3.235	3.004	3.039	4.915	5.600	4.247
-1.5	3.354	3.345	3.588	6.138	6.605	5.982
-1.7	3.377	3.180	3.780	6.275	7.543	6.885
-1.9	3.422	3.244	4.035	6.863	9.047	8.828
-2.1	3.312	3.042	4.273	8.023	10.050	10.072

Table C.3.: Calculated variance in measurement of T_{40} based on the meteorological conditions of both measurements. Deviations are listed depending on the temperature difference.

| $\Delta\Theta$ | 125 Hz | 250 Hz | 500 Hz | 1000 Hz | 2000 Hz | 4000 Hz |
°C	dB	dB	dB	dB	dB	dB
-0.1	0.01	0.00	0.00	0.01	0.03	0.09
-0.3	0.00	0.01	0.00	0.03	0.02	0.20
-0.5	0.01	0.02	0.01	0.07	0.03	0.47
-0.7	0.01	0.03	0.01	0.09	0.04	0.64
-0.9	0.01	0.02	0.01	0.11	0.10	0.67
-1.1	0.01	0.03	0.02	0.13	0.15	0.76
-1.3	0.01	0.04	0.03	0.18	0.22	0.97
-1.5	0.02	0.04	0.03	0.19	0.30	1.25
-1.7	0.02	0.04	0.03	0.22	0.38	1.52
-1.9	0.02	0.05	0.04	0.25	0.60	2.40
-2.1	0.02	0.03	0.03	0.17	0.79	3.15

Table C.4.: Calculated theoretical change of C_{80} based on the meteorological conditions of both measurements. Deviations are listed depending on the temperature difference.

| $\Delta\Theta$ | 125 Hz | 250 Hz | 500 Hz | 1000 Hz | 2000 Hz | 4000 Hz |
°C	dB	dB	dB	dB	dB	dB
-0.1	0.23	0.54	0.40	1.20	0.42	0.57
-0.3	0.22	0.23	0.15	0.33	0.28	0.36
-0.5	0.14	0.19	0.17	0.29	0.30	0.49
-0.7	0.23	0.25	0.24	0.48	0.42	0.74
-0.9	0.23	0.44	0.27	1.02	0.43	0.70
-1.1	0.24	0.44	0.32	1.03	0.41	0.72
-1.3	0.22	0.31	0.36	0.61	0.45	0.76
-1.5	0.25	0.42	0.37	0.97	0.46	0.75
-1.7	0.26	0.41	0.40	0.94	0.49	0.76
-1.9	0.27	0.42	0.40	0.91	0.56	0.93
-2.1	0.26	0.39	0.43	0.88	0.64	1.09

Table C.5.: Calculated variance in measurement of C_{80} based on the meteorological conditions of both measurements. Deviations are listed depending on the temperature difference.

Bisher erschienene Bände der Reihe
Aachener Beiträge zur Technischen Akustik

ISSN 1866-3052

Alle erschienenen Bücher können unter der angegebenen ISBN-Nummer direkt online
(http://www.logos-verlag.de) oder per Fax (030 - 42 85 10 92) beim Logos Verlag
Berlin bestellt werden.